Reactive Oxygen Species - Advances and Developments

Edited by Rizwan Ahmad

Published in London, United Kingdom

Reactive Oxygen Species - Advances and Developments
http://dx.doi.org/10.5772/intechopen.107774
Edited by Rizwan Ahmad

Contributors
José Manuel Pérez De La Lastra, Celia Andrés Juan, Celia María Curiesses Andrés, Francisco J. Plou, Eduardo Pérez-Lebeña, Alejandro Ferraz do Prado, Stefanne de Cássia Pereira da Silva, Keuri Eleutério Rodrigues, Marika Cordaro, Ramona D'Amico, Rosanna Di Paola, Salvatore Cuzzocrea, Zahid Naseer, Mudussar Nawaz, Ejaz Ahmad, Zia ur Rehman, Thérèse Hunter, Gary James Hunter, Brandon Charles Seychell, Marita Vella, Amala Reddy, Venkatramn Sairam, Tharun Srinivasan, Pavithra Ashok, Panchashree Das, Priyabrata Sen, Rizwan Ahmad

First published in London, United Kingdom, 2024 by IntechOpen
IntechOpen is the global imprint of INTECHOPEN LIMITED, registered in England and Wales, registration number: 11086078, 5 Princes Gate Court, London, SW7 2QJ, United Kingdom
Printed in Croatia

British Library Cataloguing-in-Publication Data
A catalogue record for this book is available from the British Library

Additional hard and PDF copies can be obtained from orders@intechopen.com

Reactive Oxygen Species - Advances and Developments
Edited by Rizwan Ahmad
p. cm.

This title is part of the Biochemistry Book Series, Volume 50
Topic: Cell and Molecular Biology
Series Editor: Miroslav Blumenberg
Topic Editor: Rosa María Martínez-Espinosa

Print ISBN 978-1-83768-209-6
Online ISBN 978-1-83768-253-9
eBook (PDF) ISBN 978-1-83768-254-6
ISSN 2632-0983

For EU product safety concerns:
IN TECH d.o.o., Prolaz Marije Krucifikse Kozulić 3, 51000 Rijeka, Croatia,
info@intechopen.com or visit our website at intechopen.com.

IntechOpen Book Series

Biochemistry

Volume 50

Aims and Scope of the Series

Biochemistry, the study of chemical transformations occurring within living organisms, impacts all of the life sciences, from molecular crystallography and genetics, to ecology, medicine and population biology. Biochemistry studies macromolecules - proteins, nucleic acids, carbohydrates and lipids –their building blocks, structures, functions and interactions. Much of biochemistry is devoted to enzymes, proteins that catalyze chemical reactions, enzyme structures, mechanisms of action and their roles within cells. Biochemistry also studies small signaling molecules, coenzymes, inhibitors, vitamins and hormones, which play roles in the life process. Biochemical experimentation, besides coopting the methods of classical chemistry, e.g., chromatography, adopted new techniques, e.g., X-ray diffraction, electron microscopy, NMR, radioisotopes, and developed sophisticated microbial genetic tools, e.g., auxotroph mutants and their revertants, fermentation, etc. More recently, biochemistry embraced the 'big data' omics systems. Initial biochemical studies have been exclusively analytic: dissecting, purifying and examining individual components of a biological system; in exemplary words of Efraim Racker, (1913 –1991) "Don't waste clean thinking on dirty enzymes." Today, however, biochemistry is becoming more agglomerative and comprehensive, setting out to integrate and describe fully a particular biological system. The 'big data' metabolomics can define the complement of small molecules, e.g., in a soil or biofilm sample; proteomics can distinguish all the proteins comprising e.g., serum; metagenomics can identify all the genes in a complex environment e.g., the bovine rumen.
This Biochemistry Series will address both the current research on biomolecules, and the emerging trends with great promise.

Meet the Series Editor

Miroslav Blumenberg, Ph.D., was born in Subotica and received his BSc in Belgrade, Yugoslavia. He completed his Ph.D. at MIT in Organic Chemistry; he followed up his Ph.D. with two postdoctoral study periods at Stanford University. Since 1983, he has been a faculty member of the RO Perelman Department of Dermatology, NYU School of Medicine, where he is codirector of a training grant in cutaneous biology. Dr. Blumenberg's research is focused on the epidermis, expression of keratin genes, transcription profiling, keratinocyte differentiation, inflammatory diseases and cancers, and most recently the effects of the microbiome on the skin. He has published more than 100 peer-reviewed research articles and graduated numerous Ph.D. and postdoctoral students.

Meet the Volume Editor

Dr. Rizwan Ahmad is a University Professor and the Coordinator of the PG Accreditation College of Medicine at Imam Abdulrahman bin Faisal University in KSA. Previously, he worked as an Associate Professor of Human Function at Oman Medical College in Oman and SBS University in India. Dr. Ahmad completed his education at Aligarh Muslim University in India and received a fellowship from the Indian Council of Medical Research. He has authored several articles in peer-reviewed journals, edited many books, and served as a reviewer for various journals.

Contents

Preface **XIII**

Chapter 1 **1**
Introductory Chapter: Reactive Oxygen Species – Origin and Significance
by Rizwan Ahmad

Chapter 2 **9**
Reactivity and Applications of Singlet Oxygen Molecule
by Celia María Curieses Andrés, José Manuel Pérez de la Lastra, Celia Andrés Juan, Francisco J. Plou and Eduardo Pérez-Lebeña

Chapter 3 **33**
Relevance of Oxidoreductases in Cellular Metabolism and Defence
by Panchashree Das and Priyabrata Sen

Chapter 4 **51**
The Good and the Bad: The Bifunctional Enzyme Xanthine Oxidoreductase in the Production of Reactive Oxygen Species
by Brandon Charles Seychell, Marita Vella, Gary James Hunter and Thérèse Hunter

Chapter 5 **71**
Methylmercury Promotes Oxidative Stress and Activation of Matrix Metalloproteinases: Cardiovascular Implications
by Keuri Eleutério Rodrigues, Stefanne de Cássia Pereira da Silva and Alejandro Ferraz do Prado

Chapter 6 **91**
Oxidative Stress and Reproduction Health: Physiology, Pathology, and Clinical Biomarkers
by Ramona D'Amico, Salvatore Cuzzocrea, Marika Cordaro and Rosanna Di Paola

Chapter 7 **115**
Oxidative Stress and Male Fertility: Promising Role of Nutraceuticals
by Zahid Naseer, Mudussar Nawaz, Ejaz Ahmad and Zia ur Rehman

Chapter 8 **141**
Regulating Reactive Oxygen Species in Rheumatoid Arthritis: Insights into Cell Signaling Pathways and Nano-Particles as Carriers
by Tharun Srinivasan, Pavithra Ashok, Venkatraman Sairam and Amala Reddy

Preface

The term "Reactive Oxygen Species" or ROS is used to describe the highly active radicals formed by oxygen's unpaired electrons, such as the hydroxyl radical (•OH) and superoxide (•O2−). ROS also includes nonradical oxidizing agents like singlet oxygen ($^1O_2$), ozone (O_3), hydrogen peroxide (H_2O_2), and hypochlorous acid (HOCl). These free radicals have been recognized in chemistry since the early 20th century and have been found in biological systems since 1954.

The first chapter discusses the origin and significance of ROS. ROS play a crucial role in cell physiology. At low to moderate levels, they activate cell proliferation, migration, invasion, and angiogenesis. However, high levels of ROS can cause significant damage to various cellular components, such as proteins, nucleic acids, lipids, membranes, and organelles, ultimately leading to cell death. Recent studies indicate that targeting ROS levels can effectively treat cancer, including immunotherapies with positive *in vitro* and *in vivo* results.

Singlet oxygen, although the lowest energy-excited state of molecular oxygen, is a highly reactive species that initiates oxidation reactions of biomolecules such as amino acids, proteins, nucleic acids, and lipids. This happens either by a direct response or by the induction of ROS. Singlet oxygen is a highly reactive electrophilic species that reacts with electron-rich molecules and is strongly linked to several pathologies. The second chapter provides information on the physicochemical properties of a molecule, including methods of generation and detection, as well as its reactivity. The chapter also discusses Photodynamic therapy (PDT) as a potential treatment for cancer and other diseases. PDT is a non-invasive, selective, and localized method of destroying tumor cells with fewer side effects.

Oxidoreductases are enzymes found in various living organisms such as plants, animals, and microorganisms. In plant biochemistry, oxidoreductases refer to a large group of enzymes having EC 1 nomenclature, which catalyze oxidation and reduction reactions in plants. These enzymes mainly facilitate the transfer of electron molecules from one molecule (known as electron donor or reductant) to another (known as electron acceptor or oxidant). Plant plasma membranes owe a great deal of their functionality to oxidoreductase enzymes. These enzymes diligently maintain the redox potential of the membranes through a variety of mechanisms such as proton pumping, regulation of nutrient uptake, ion channel regulation, and iron reduction. Furthermore, they play an important role in signal transduction pathways and help defend against pathogen attack. Overall, the third chapter on oxidoreductases in plant plasma membranes is crucial to plant biology, ensuring proper functioning and defense against external threats.

Xanthine oxidoreductase (XOR) is an enzyme that produces reactive oxygen species (ROS). While ROS generated by XOR can help protect against infection, they can also contribute to inflammation, oncogenesis, brain injury, and stroke. XOR is involved in various conditions, such as tumor lysis syndrome in chemotherapy patients and ischemia-reperfusion injury, which can increase ROS levels in the body. The fourth chapter on XOR summarizes that its blood presence can serve as a biomarker for diagnosing various conditions, including cardiovascular disease and oxidative stress.

In the chapter 5, it is suggested that predatory aquatic mammals are the primary source of Methylmercury, which can lead to poisoning and irreversible damage to various organs. The evidence gathered so far shows that MeHg tends to alter the redox state by increasing the production of reactive species. Moreover, MeHg reduces ROS inactivation by depleting GSH levels, inactivating thiol enzymes, and enzymes such as SOD and catalase. These findings can contribute significantly to the understanding of MeHg's impact on the environment and human health.

Chapter 6 on oxidative stress and reproductive health sheds light on the crucial role of reactive oxygen species (ROS) in the physiological processes and signaling pathways related to male and female fertility. In females, oxidative stress is involved in critical ovarian functions, cyclical changes in the endometrium, embryo development, tubal operations, pregnancy, and complications such as miscarriages, recurrent pregnancy losses, preeclampsia, and gestational diabetes. High concentrations of ROS can lead to sperm pathologies, causing ATP depletion and inadequate axonemal phosphorylation or lipid peroxidation, which in turn can lead to a loss of sperm motility and viability. This chapter will discuss ROS's mechanisms, production, physiological and pathophysiological roles in the male and female reproductive system and the latest advances in diagnostic methods that use ROS as biochemical markers.

It's important to note that oxidative stress can hurt male fertility. However, many biological agents, such as minerals, vitamins and herbs, can have a positive effect. These agents contain active compounds that can help improve enzymatic antioxidant activity and indirectly enhance antioxidant levels in the body, which can help clear excessive ROS from the male reproductive system. This can lead to an improvement in libido and penile erection, which is significant. In chapter 7, the authors have explained that different antioxidative mechanisms associated with well-known nutraceutical compounds improve other aspects of sperm health, such as spermatogenesis, motility, concentration, morphology, mitochondrial function, and DNA integrity of sperm.

The last chapter focuses on the relationship between ROS and rheumatoid arthritis (RA), underlining the crucial role of cell signaling pathways in controlling ROS. RA is a chronic, debilitating inflammatory condition characterized by the degradation of joints and permanent disability. This autoimmune disease affects millions of individuals worldwide, and can cause significant pain, swelling, and stiffness in the joints. Excessive production of ROS is undeniably a significant factor in the development of RA, causing tissue damage and oxidative stress. In recent years, researchers have been exploring the potential of nano-particles as carriers for ROS regulation therapies in RA treatment, with promising results. The potential of nano-particles as targeted drug delivery systems for scavenging excess ROS and restoring redox equilibrium within affected cells is a promising avenue for RA treatment and deserves further exploration.

In the past few decades, significant progress has been made in understanding the sources of ROS production, the mechanisms of ROS regulation, and the implications of ROS in disease pathogenesis. Moreover, novel therapeutic strategies targeting ROS have emerged, including antioxidants, scavengers, and modulators of ROS signaling pathways. *Reactive Oxygen Species – Advances and Developments* is a comprehensive book that delves into the intricacies of ROS, providing invaluable insights to researchers in the field. By providing a comprehensive overview of the subject matter, the book equips researchers with the essential tools and knowledge to advance their work in this critical area.

I extend my sincere gratitude to Dr. Bassam Awary and Professor Mahdi Abumadni for their constant support. I would like to express my profound appreciation to my family members for their unwavering encouragement and assistance throughout the book project. Furthermore, I would like to thank Ms. Ana Cink, publishing process manager, and the team at In Tech Publishers UK for their prompt support and guidance.

Dr. Rizwan Ahmad
College of Medicine,
Imam Abdulrahman Bin Faisal University,
Dammam, Saudi Arabia

Chapter 1

Introductory Chapter: Reactive Oxygen Species – Origin and Significance

Rizwan Ahmad

1. Introduction

In 1900, while serving as a chemistry faculty at Michigan State University, Professor Moses Gomberg made a significant discovery that revolutionized the field of chemistry: the free radical. This concept was later recognized as a National Historic Chemical Landmark by the American Chemical Society in 2000, acknowledging the importance of his work. In 1954, Gershman proposed the "free radical theory of oxygen toxicity." This theory suggests that oxygen is toxic because it can form free radicals. Another chemist, Fenton, discovered that the reaction between hydrogen peroxide and ferrous sulfate produces a violet color, which is the oxidation of tartaric acid upon adding alkali. This reaction is now known as the Fenton reaction and serves as the basis for producing the hydroxyl radical. These discoveries have paved the way for further exploration into free radical science and contributed significantly to our understanding of chemical reactions and their implications [1–4].

Free radicals are molecules that contain one or more unpaired electrons and are capable of independent existence. They are highly reactive and play a crucial role in various metabolic processes, such as oxidative reactions in mitochondria and oxidative bursts of phagocytes. However, in excess, free radicals can cause diseases such as autoimmune, cardiovascular, neurodegenerative conditions, and cancer. Reducing the number of free radicals is necessary to minimize these pathological conditions.

Reactive oxygen species (ROS) and reactive nitrogen species (RNS) are two types of free radicals that are formed in the body and consist of both radical and nonradical moieties [1]. It is important to note that the term "ROS" refers to a group of small molecules of oxidizing, nitrosating, nitrating, halogenating, and thiol-reactive species produced in biological systems. This group does not represent a single species, making it nonspecific and ambiguous. Some experts have criticized using ROS as an umbrella term for oxidants because of its lack of specificity. In 2014, Holmstrom and Finkel clarified the dual nature of reactive oxygen species (ROS), their role in cell damage and protection, and their involvement in various diseases. "*Nonetheless, from a biological point of view, it is beginning to look as if ROS are neither cellular heroes nor villains—but instead something that occupies that always entertaining, captivating, and fertile middle ground*" [5].

2. Sources of ROS

ROS is primarily expanded to include reactive oxygen-containing compounds or nonradical oxidizing agents such as singlet oxygen ($^1O_2$), ozone (O_3), hydrogen peroxide (H_2O_2), and hypochlorous acid (HOCl). ROS can originate from various sources, such as the photolysis of gaseous ozone, materials-mediated catalytic reactions, and endogenous activities in biological systems [6]. Researchers have magnified their focus on studying ROS in living organisms in recent years because ROS plays an indisputable role in regulating various physiological functions. Living organisms have two sources of ROS: exogenous and endogenous. Exogenous sources include engineered nanoparticles (NPs) exposure, radiation, chemotherapeutics, and microbial infection. It is crucial to note that the reactive oxygen species (ROS) acts as a byproduct during cellular respiration and normal metabolic processes [6–9]. High-concentration ROS are highly toxic to living organisms, but in normal physiological processes, ROS act as messengers in various cellular functions and can be identified as signal molecules or regulators in living systems [10]. It is believed that the effect of ROS on physiological processes is attributed to their ability to alter specific proteins' activity [11].

In recent decades, ROS have been studied extensively for their roles in normal physiological processes. These roles include blood vessel modulation, immune function, oxygen sensing, gene activation, and cellular growth [12, 13]. ROS plays a crucial role in initiating and developing various pathological processes, including aging, cancer, insulin resistance, diabetes mellitus, cardiovascular diseases, Alzheimer's disease, and more [7, 14]. Excessive ROS generation can lead to tissue dysfunction or cell death, while a stable concentration of ROS can effectively regulate physiological processes. Therefore, maintaining redox homeostasis is crucial to keeping normal physiological functions and reducing the incidence of diseases [7].

Antioxidants like vitamin E (tocopherols and tocotrienols), SOD, CAT, GPx, vitamin C (ascorbic acid), beta-carotene (β-carotene), and coenzyme Q10 (CoQ10) play a crucial role in neutralizing and eliminating ROS. They are essential to the body's defense against these toxic species. Polyphenolic compounds, another type of antioxidant, can also reduce the levels of harmful ROS by scavenging them, which helps prevent oxidative damage to macromolecules [4, 15].

3. Oxidative stress and diseases

Oxidative stress (OS) occurs when reactive oxygen species exceed the body's capacity to neutralize them with antioxidants. This imbalance leads to the accumulation of ROS, which can cause significant harm to the body's cellular structures. OS disrupts the delicate balance between prooxidants and antioxidants, resulting in severe damage to cellular macromolecules and potentially lethal consequences [16]. Various methods have been developed and employed to measure the nature and extent of oxidative stress, ranging from the oxidation of lipids to free amino acids proteins and DNA. Although diverse oxidative stress biomarkers are available as predictors of various diseases, the specificity of each is yet to be established [17]. Autophagy is a remarkable catabolic process that efficiently delivers cytoplasmic macromolecules and organelles to the lysosomes for degradation [18]. Oxidative stress plays a crucial role in regulating the process of autophagic flux by influencing the transcription of autophagic genes, the activity of proteins, and the degradation of organelles.

During periods of oxidative stress, ROS triggers macroautophagy (commonly known as autophagy), and the selective degradation of oxidized proteins occurs through CMA (chaperone-mediated autophagy) to ensure cellular viability. The susceptibility of CMA substrates to degradation increases because of oxidative modification and LAMP2a. (lysosome-associated membrane glycoprotein two alpha) upregulation enhances CMA in cells challenged with ROS. However, the effect of ROS on micro-autophagy remains unclear, and further research is needed to understand the roles of macroautophagy, microautophagy, and CMA in response to oxidative stress [18, 19].

The over-secretion of ROS in the brain leads to oxidative stress that, if not suppressed or inhibited, could lead to oxidative damage of essential components of the central nervous system. ROS can also initiate or enhance some reactions that may have detrimental effects on the physiological functions and health of the brain [20]. If not abated, these reactions, such as neuroinflammation and progressive neuronal cell loss *via* apoptosis, can exacerbate protein misfolding and formation of protein aggregates, resulting in neurodegeneration and associated neurobehavioral incompetence. Considering the pivotal roles of oxidative stress, neuroinflammation, protein misfolding, and apoptosis in neurodegenerative diseases, the manipulation of these significant players in each of the pathological mechanisms may represent a promising treatment option to slow down neurodegeneration and alleviate associated symptoms [20].

Autoimmune diseases, systemic lupus erythematosus (SLE), and rheumatoid arthritis (RA) are notable examples in which free radicals cause damage to cells and tissues. Arthritic inflammation is a medical condition where ROS acts as a potent inflammatory mediator, contributing significantly to the destruction of collagen tissues. The overproduction of ROS not only causes damage leading to the breakdown of cartilage but also triggers the apoptosis of the vital chondrocytes responsible for the structure and function of the cartilage. The reduced number of chondrocytes, in turn, results in an inability of the cartilage to self-repair, promoting the breakdown of the extracellular matrix in joints. Moreover, the H_2O_2 molecule causes chondrocyte lipid peroxidation, significantly affecting protein oxidation and cartilage matrix degradation. Scientific studies have established the crucial role of oxidative stress in joint pathology, which includes inflammatory infiltration, synovial proliferation, and angiogenesis. These findings could pave the way for more targeted and effective treatments for these diseases in the future [21–24].

Diabetes mellitus is a chronic condition that results in high levels of blood glucose, known as hyperglycemia. This condition can be caused by a defect in insulin secretion (in Type I diabetes), resistance to insulin action (in Type II diabetes), or both [24]. Common symptoms of diabetes include thirst, hunger, weight loss, and weakness, which can eventually lead to coma. Diabetes is often linked to increased free radicals or decreased antioxidant systems, leading to oxidative stress development [25]. Both mitochondrial and non-mitochondrial sources cause DM *via* reactive oxygen species. Superoxide is mainly produced by electron transport chain complexes I and III under normal conditions. Increased glucose levels in diabetes mellitus led to elevated pyruvate production through increased glycolysis. It causes an increase in the inner mitochondrial membrane potential, leading to mitochondrial dysfunction and ROS production at the electron transport chain complex II. By understanding these mechanisms, researchers can work toward developing effective strategies to mitigate the impact of oxidative stress on patients with DM [26, 27].

It is worth noting that DNA mutation, often caused by oxidative damage, plays a vital role in cancer development. Cancer cells, in particular, in comparison to normal cells, have higher levels of ROS and are more susceptible to mitochondrial dysfunction because of their higher metabolic rate [28]. Cancer cells display elevated levels of oxidative stress due to the activation of oncogenes and loss of tumor suppressors [29].

ROS alters the growth signals and gene expression and causes continuous proliferation of cancer cells [30]. ROS can damage DNA by inducing base modifications, deletions, strand breakage, chromosomal rearrangements, and hyper- and hypo-methylation of DNA. It is one of the initiators of carcinogenesis, wherein many tumors exhibit elevated levels of DNA damage. Epidemiological studies show an increased cancer risk with low antioxidant levels. Patients with various cancers exhibit increased total oxidant status levels and decreased total antioxidant status [31].

Cardiovascular diseases (CVDs) are complex medical conditions affecting the heart and blood vessels. It is important to note that atherosclerosis, also known as hardening of the arteries, is caused by high levels of lipids in the blood, a significant risk factor. Patients with atherosclerosis, Type 2 diabetes, and obesity often have elevated oxidized low-density lipoprotein (LDL), glucose, and free fatty acids [31, 32]. Atherosclerosis occurs due to an imbalance between oxidants and antioxidants, leading to oxidative stress. Cells in the vessel wall, including endothelial cells, smooth muscle cells, and macrophages, produce free radicals contributing to this imbalance. This results in endothelial dysfunction, leading to increased endothelial permeability, upregulation of endothelial adhesion molecules, and inflammatory cell infiltration into the arterial wall [32]. ROS (reactive oxygen species) plays a significant role in endothelial injury, dysfunction, and lesion progression. Matrix metalloproteinases (MMPs) are activated by ROS, leading to intimal extracellular matrix degradation and smooth muscle cell migration. Cigarette smoking contains a large number of free radicals. It may downregulate essential exogenous and endogenous antioxidants such as vitamin D, carotenes, GPx, and SOD, leading to the dysfunction of monocytes and vascular smooth muscle cells. Proatherogenic agents such as oxidized lipids, high glucose, and cigarette constituents increase free radical production [33, 34].

Hypertension affects 40% of adults worldwide, making it a significant health issue. Free radical-induced oxidative stress, in part, contributes to endothelial dysfunction and the development of hypertension [35]. One of the causes of hypertension is the increased generation of ROS, which decreases the bioavailability of nitric oxide (NO•) by forming peroxynitrite (ONOO−), resulting in reduced endothelium-dependent vasodilation. Patients with hypertension present with decreased NO bioavailability and increased oxidative stress. The oxidation-induced impairment of NO also causes a reduction in its ability to counteract the vasoconstrictive and hypertensive effects of angiotensin II. Angiotensin II, in turn, promotes oxidative stress and decreases NO bioavailability [36, 37].

4. ROS as therapeutic agents

Over the past few decades, ROS have emerged as potential therapeutic products, finding applications in various areas, from wound healing and hair growth enhancement to cancer treatment, stem cell differentiation, and tissue engineering. Despite significant advances, several challenges still need to be overcome for the clinical translation of ROS. Several studies have found that ROS can benefit at low concentrations and may be a cost-effective and convenient way to induce tissue regeneration [38, 39].

Recent research has revealed that cancer cells are more sensitive to oxidative damage, which has led to the development of methods for selectively killing tumor cells using ROS. As a result, there is extensive investigation into creating sensors to monitor ROS production during cancer treatment. The high oxidative status of cancer cells led to the development of H_2O_2-responsive anticancer prodrugs. Several drugs inducing oxidative stress are under development to specifically target the susceptibility of cancer cells [38, 40]. The precise control of stem cell differentiation by ROS through ROS stimulation techniques such as cold atmospheric plasma and using biomaterials that release ROS has shown potential for tissue engineering applications. Some studies have demonstrated *in vitro* neurogenesis through cold atmospheric plasma stimulation techniques, whereas others are exploring using different biomaterials that release ROS [41]. Nano-catalytic medicines have emerged as a promising approach for cancer treatment by producing lethal ROS within tumors. This innovative approach has garnered significant attention lately, as it offers a tumor-specific treatment that could potentially eradicate cancer cells [42]. However, further research is required to fully understand the possible side effects of these techniques.

5. Conclusion

Reactive oxygen species (ROS) are unique molecules produced naturally through cell metabolic reactions. These molecules play a crucial role in cell function and survival by acting as signaling agents that regulate specific biochemical pathways. However, when there is an imbalance in ROS signaling or excessive ROS production, it can adversely impact disease pathophysiology. Therefore, an in-depth understanding of tissue-specific redox signaling complexities is essential to develop new and innovative therapies for cardiovascular and metabolic disease pathogenesis. Despite the potential benefits of ROS in various fields of biology, such as cancer treatment, tissue engineering, wound healing, and developmental biology research, they possess a complex mechanism of action with possible side effects. It is crucial to develop new systems that can regulate the timing of ROS production and control their levels. Further research and advancements in this field will pave the way for a better understanding of ROS and their potential applications in various fields of biology.

Author details

Rizwan Ahmad
College of Medicine, Imam Abdulrahman bin Faisal University, Dammam, Saudi Arabia

*Address all correspondence to: ahmadriz.biochem@gmail.com

References

[1] Ahmad R. In: Ahmad R, editor. Introductory Chapter: Basics of Free Radicals and Antioxidants, Free Radicals, Antioxidants and Diseases. Rijeka: Intech Open; 2018

[2] American Chemical Society National Historic Chemical Landmarks. Moses Gomberg and Organic Free Radicals. 2000. Available from: http://www.acs.org/content/acs/en/education/whatischemistry/landmarks/freeradicals

[3] Fenton HJH. Oxidation of tartaric acid in presence of iron. Journal of the Chemical Society. 1894;**65**:899-910

[4] Gerschman R, Gilbert DL, Nye SW, Dwyer P, Fenn WO. Oxygen poisoning and x-irradiation: A mechanism in common. Science. 1954;**119**:623-626

[5] Ahmad R, Ahsan H. Singlet oxygen species and systemic lupus erythematosus: A brief review. Journal of Immunoassay & Immunochemistry. 2019;**40**(4):343-349

[6] Holmström KM, Finkel T. Cellular mechanisms and physiological consequences of redox-dependent signalling. Nature Reviews. Molecular Cell Biology. 2014;**15**(6):411-421

[7] Bayir H. Reactive oxygen species. Critical Care Medicine. 2005;**33** (12 Suppl):498-501

[8] Spooner R, Yilmaz O. The role of reactive-oxygen-species in microbial persistence and inflammation. International Journal of Molecular Sciences. 2011;**12**(1):334-352

[9] Zhang C, Wang X, Du J, Gu Z, Zhao Y. Reactive oxygen species-regulating strategies based on nanomaterials for disease treatment. Advanced Science. 2021;**8**(3):2002797

[10] Manke A, Wang L, Rojanasakul Y. Mechanisms of nanoparticle-induced oxidative stress and toxicity. BioMed Research International. 2013;**2013**:942916

[11] Circu ML, Aw TY. Reactive oxygen species, cellular redox systems, and apoptosis. Free Radical Biology Medicine. 2010;**48**(6):749-762

[12] Pourova J, Kottova M, Voprsalova M, Pour M. Reactive oxygen and nitrogen species in normal physiological processes. Acta Physiologica (Oxford, England). 2010;**198**(1):15-35

[13] Di Meo S, Reed TT, Venditti P, Victor VM. Role of ROS and RNS sources in physiological and pathological conditions. Oxidative Medicine and Cellular Longevity. 2016;**2016**:1-44

[14] Alfadda AA, Sallam RM. Reactive oxygen species in health and disease. Journal of Biomedicine & Biotechnology. 2012;**2012**:936486

[15] Nash KM, Ahmed S. Nanomedicine in the ROS-mediated pathophysiology: Applications and clinical advances. Nanomedicine. 2015;**11**(8):2033-2040

[16] Ahsan H, Yusuf Hasan M, Ahmad R. Reactive oxygen species (ROS) in the pathophysiology of rheumatoid arthritis (RA). In: Ahmad R, editor. Reactive Oxygen Species. Rijeka: Intech Open; 2022

[17] Akhigbe R, Ajayi A. The impact of reactive oxygen species in the development of cardiometabolic disorders: A review. Lipids in Health and Disease. 2021;**20**:23

[18] Fridovich I. Fundamental aspects of reactive oxygen species, or what's the matter with oxygen? Annals of the New York Academy of Sciences. 1999;**893**(1):13-18

[19] Kiffin R, Christian C, Knecht E, Cuervo AM. Activation of chaperone-mediated autophagy during oxidative stress. Molecular Biology of the Cell. 2004;**10**(15):4829-4840

[20] Bento CF, Renna M, Ghislat G, Puri C, Ashkenazi A, Vicinanza M, et al. Mammalian autophagy: How does it work? Annual Review of Biochemistry. 2016;**85**:685-713

[21] Olaleye Oladele J, Oladiji A, Titilope Oladele O, et al. Reactive oxygen species in neurodegenerative diseases: Implications in pathogenesis and treatment strategies. In: Ahmad R, editor. Reactive Oxygen Species. London, UK, Rijeka: Intech Open; 2022. p. 2022

[22] Ahmad R, Hussain A, Ahsan H. Peroxynitrite: Cellular pathology and implications in autoimmunity. Journal of Immunoassay & Immunochemistry. 2019;**40**:123-138

[23] Balogh E, Veale DJ, McGarry T, Orr C, Szekanecz Z, Ng CT, et al. Oxidative stress impairs energy metabolism in primary cells and synovial tissue of patients with rheumatoid arthritis. Arthritis Research & Therapy. 2018;**20**(1):95

[24] Mateen S, Moin S, Zafar A, Khan AQ. Redox signaling in rheumatoid arthritis and the preventive role of polyphenols. Clinica Chimica Acta. 2016;**463**:4-10

[25] Ahmad R, Ahsan H. Dual autoimmune diseases: Rheumatoid arthritis with systemic lupus erythematosus and type 1 diabetes mellitus with multiple sclerosis. Rheumatology & Autoimmunity. 2022;**2**(3):120-128

[26] Gavin JR, Alberti KGMM, Davidson MB, DeFronzo RA, Drash A, Gabbe SG, et al. Report of the expert committee on the diagnosis and classification of diabetes mellitus. Diabetes Care. 1997;**20**:1183-1197

[27] Oberley LW. Free radicals and diabetes. Free Radical Biology & Medicine. 1988;**5**(2):113-124

[28] Kwong LK, Sohal RS. Substrate and site specificity of hydrogen peroxide generation in mouse mitochondria. Archives of Biochemistry and Biophysics. 1998;**350**(1):118-126

[29] Bajaj S, Khan A. Antioxidants and diabetes. Indian Journal of Endocrinology Metabolism. 2012;**2**:S267-S271

[30] Acuna UM, Wittwer J, Ayers S, Pearce CJ, Oberlies NH, De Blanco EJ. Effects of (5Z)-7-oxozeaenol on the oxidative pathway of cancer cells. Anticancer Research. 2012;**32**(7):2665-2671

[31] Cairns RA, Harris I, McCracken S, Mak TW. Cancer cell metabolism. Cold Spring Harbor Symposia on Quantitative Biology. 2011;**76**:299-311

[32] Valko M, Leibfritz D, Moncol J, Cronin MT, Mazur M, Telser J. Free radicals and antioxidants in normal histological functions and human disease. The International Journal of Biochemistry & Cell Biology. 2007;**39**(1):44-84

[33] Phaniendra A, Jestadi DB, Periyasamy L. Free radicals: Properties, sources, targets, and their implication in various diseases. Indian Journal of Clinical Biochemistry. 2014;**30**(1):11-26

[34] Ross R. Atherosclerosis—An inflammatory disease. The New England Journal of Medicine. 1999;**340**(2):115-126

[35] Rajagopalan S, Meng XP, Ramasamy S, Harrison DG, Galis ZS. Reactive oxygen species produced by macrophage-derived foam cells regulate the activity of vascular matrix metalloproteinases *in vitro*. Implications for atherosclerotic plaque stability. The Journal of Clinical Investigation. 1996;**98**(11):2572-2579

[36] Barnoya J, Glantz SA. Cardiovascular effects of secondhand smoke: Nearly as large as smoking. Circulation. 2005;**111**(20):2684-2698

[37] WHO. A Global Brief on Hypertension. World Health Day. Geneva, Switzerland: WHO Press; 2013. WHO Reference number: WHO/DCO/WHD/2013.2

[38] Touyz RM. Reactive oxygen species, vascular oxidative stress, and redox signaling in hypertension: What is the clinical significance? Hypertension. 2004;**44**(3):248-252

[39] Touyz RM. Reactive oxygen species and angiotensin II signaling in vascular cells: Implications in cardiovascular disease. Brazilian Journal of Medical and Biological Research. 2004;**37**(8):1263-1273

[40] Graceffa V. Therapeutic potential of reactive oxygen species: State of the art and recent advances. Slas Technology/Translating Life Sciences Innovation. 2020;**26**(2):40-158; Sage Publishing, US

[41] Nakamura H, Takada K. Reactive oxygen species in cancer: Current findings and future directions. Cancer Science. 2021;**112**(10):3945-3952

[42] Su G, Xu H, Zhou FF, Gong X, Tan S, He Y. Boosting reactive oxygen species generation with a dual-catalytic nanomedicine for enhanced tumor nanocatalytic therapy. ACS Applied Materials & Interfaces. 2023;**15**(51):59175-59188. DOI: 10.1021/acsami.3c13882

Chapter 2

Reactivity and Applications of Singlet Oxygen Molecule

Celia María Curieses Andrés, José Manuel Pérez de la Lastra, Celia Andrés Juan, Francisco J. Plou and Eduardo Pérez-Lebeña

Abstract

Reactive oxygen species (ROS) are molecules produced in living organisms, in the environment, and in various chemical reactions. The main species include, among others, singlet oxygen (1O2), the superoxide anion radical (•O^{2-}), the hydroxyl radical (HO•), and the hydroperoxyl radical (HOO•). In general, the reactivity of $^1O_2$ is lower than that of HO• but even higher than that of •O^{2-}. Singlet oxygen is the lowest energy excited state of molecular oxygen, but it is also a highly reactive species, which can initiate oxidation reactions of biomolecules such as amino acids, proteins, nucleic acids, and lipids, either by a direct reaction or by the induction of ROS. Singlet oxygen is a highly reactive electrophilic species that reacts with electron-rich molecules and is related to several types of pathologies. To inhibit the oxidation of biomolecules with this species, some substances act as antioxidants by performing a quenching effect. In this chapter, aspects such as its physicochemical properties, methods of generation and detection, as well as the reactivity of this molecule are detailed.

Keywords: singlet oxygen, photochemistry, phototherapy, photosensitizers, singlet oxygen generation methods

1. Introduction: Singlet oxygen ($^1O_2{}^1\Delta g$ or $^1O_2{}^*$)

Oxygen is the most abundant element in the earth's crust, mainly in its gaseous biatomic molecule form, constituting 21% by volume of dry air. It is one of the most biologically important elements because of the type and number of reactions in which it participates, providing the thermodynamic force necessary for the metabolism of all higher organisms. Molecular oxygen in its basal state has an open-shell electronic configuration, with two electrons of equal spin occupying different degenerate molecular orbital. It is only capable of accepting one electron at a time during a redox (radical-type) reaction, reacting slowly with most organic molecules. It has two excited states of singlet multiplicity. The lower energy state is designated as $O_2(^1\Delta g)$ or $^1O_2$, and the electron distribution in $^1\Delta_g$ has antiparallel spins, the two electrons occupy the same orbital with opposite spins, therefore the spin restriction does not exist and it is able to accept two electrons at a time, thus increasing its oxidative capacity compared to $^3O_2$. The next excited singlet level is represented by the symbol $O_2(^1\Sigma_g+)$, **Figure 1**.

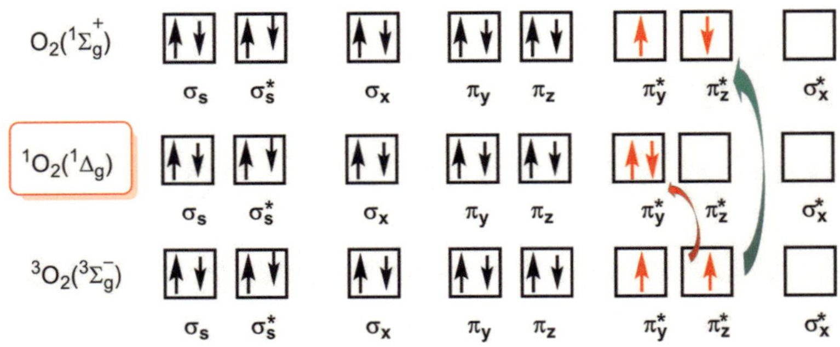

Figure 1.
Electronic configuration molecular oxygen and the first two states mentioned.

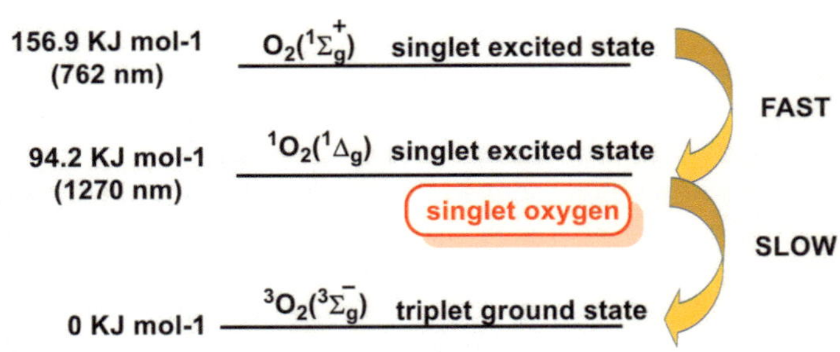

Figure 2.
Energy diagram for O_2. The singlet states of oxygen are 156.9 and 94.2 kJ/mol higher in energy than its ground triplet state.

Because of the high energy, low stability, and shorter lifetimes of the O_2 state $({}^1\Sigma_g{}^+)$, the term "singlet oxygen" commonly refers to the ${}^1O_2$ state. Singlet oxygen, or the first excited state of oxygen, is a highly oxidizing species that is generated by a light-activated compound, called a photosensitizer, and appears to play a significant role in solution reactions. It is found at 22.5 Kcal/mol above the basal state. This species emits phosphorescent light in the infrared region (1270 nm). The higher energy state $({}^1S_g{}^+)$, on the other hand, is rapidly deactivated at ${}^1D_g$ and therefore will not be as reactive. The two energetically closest electronically excited states are singlet states, whose spectroscopic notations are $1\Delta_g$ and $1\Delta_g{}^+$. Singlet oxygen $({}_1O^2)$ is an electronically excited species of molecular oxygen (O2) and participates in numerous oxidation reactions as an activated species and plays a key role in many biological processes, **Figure 2**.

The transition between the triplet ground state $({}^3\Sigma_g{}^-)$ and the first excited singlet $({}_1\Delta_g)$ is spin, symmetry, and parity forbidden; therefore, direct excitation of the ground state to form singlet oxygen is very unlikely, and gas-phase singlet oxygen has an extremely long lifetime (72 min) [1]; however, interaction with solvents reduces the lifetime, depending on the solvent and varies from 10^{-3} to 10^{-6} s [2], with the shortest lifetime observed in water [1].

From the values in **Table 1**, it can be concluded that some of the solvent characteristics that influence the lifetime are the number of C-H and O-H bonds and the presence of halogen and deuterium atoms [2].

Solvent	τ(μs)	Solvent	τ(μs)
H_2O	3.1	CH_3CN	77.1
CH_3OH	9.5	CH_2Cl_2	99
C_6H_{14}	23.4	D_2O	68
C_6H_6	30.0	C6D6	681
$(CH_3)_2CO$	51.2	$(CD_3)_2CO$	992

Table 1.
Lifetime of $^1O_2$ in different solvents.

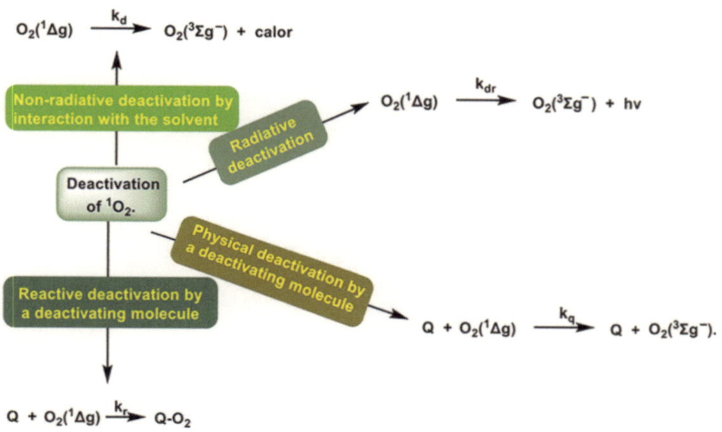

Figure 3.
Singlet oxygen deactivation pathways. The molecule that interacts with $O_2(^1\Delta_g)$ is commonly called quencher (Q).

2. Singlet oxygen chemistry

Because $^1O_2$ is a molecule in an electronically excited state, it is very unstable with respect to its ground state, and once generated it can undergo various spontaneous processes to deactivate it. Such deactivation can occur by different routes, **Figure 3**.

The non-radiative deactivation process of $^1O_2$ (kd) involves the transfer of electronic energy to the vibrational levels of the solvent. The radioactive deactivation process of $^1O_2$ (kdr) occurs when $^1O_2$ decays to its ground state emitting phosphorescent radiation of 1270 nm. The value of this rate constant depends on the medium.

The deactivation processes by molecules in the medium can be either physical (kq) or chemical (kr). Physical deactivation of $^1O_2$ occurs mainly by energy transfer mechanisms. Molecules with several conjugated double bonds, such as long-chain polyenes, quinones, dyes, and transition metal complexes, are some examples where this type of deactivation occurs [3].

The reactive deactivation of singlet oxygen, as the name suggests, refers to its chemical reaction with a wide variety of organic molecules. Singlet oxygen is a highly reactive electrophilic species and has the ability to rapidly attack organic compounds. It is about 1000 times more reactive than the basal state of oxygen.

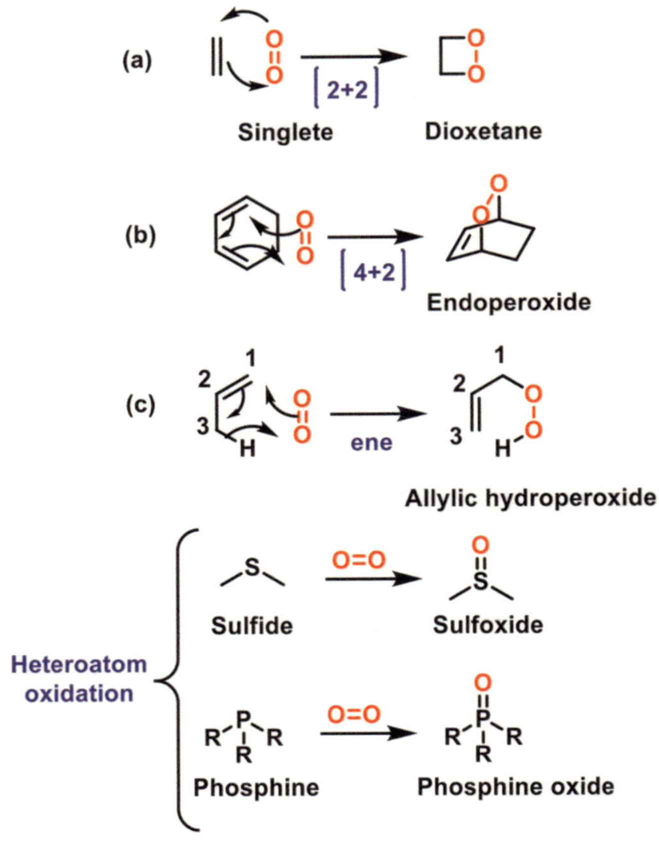

Figure 4.
Singlet oxygen-mediated photooxidations: (a) cycloaddition [2 + 2], (b) [4 + 2] cyclo addition, (c) 1,3-ene addition, and hetero atom oxidation. In all these reactions the primary products formed can undergo rearrangements to give a wide range of stable oxidized products.

This high reactivity is simply because many of the substances with which it reacts are in the singlet basal state, so the reaction is a singlet-singlet reaction, more likely than a triplet-singlet reaction, as it should be with oxygen in its basal state. As singlet molecular oxygen possesses a pair of electrons with opposite spins in the highest occupied molecular orbital, they give $O_2(^1\Delta_g)$ dienophilic properties, which explains its significant reactivity towards electron-rich organic molecules, particularly those with conjugated double bonds [4]. These reactions are electrophilic [4π + 2π], [2π + 2π], and ene-type additions, leading to the formation of allylic hydroperoxides, dioxetanes, or endoperoxides [5–8].

Singlet oxygen also oxidizes sulfur, selenium, phosphorus, and nitrogen compounds, **Figure 4**. Singlet oxygen-mediated oxidation reactions have been explored with a variety of organic molecules containing heteroatoms.

3. Singlet oxygen production

There are three main sources for generating singlet oxygen: photochemical, chemical, and *in vivo* reactions, **Figure 5**.

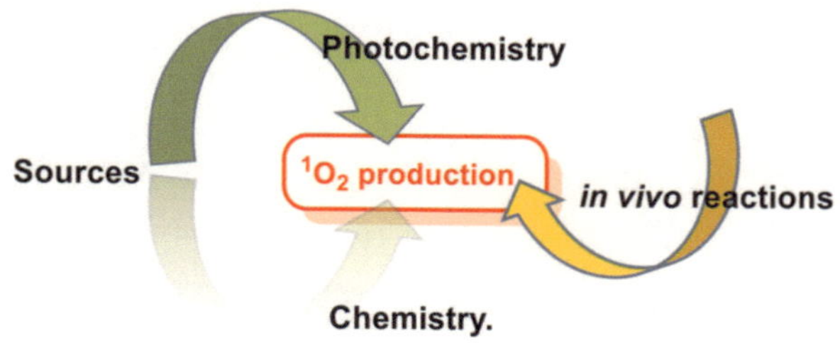

Figure 5.
Sources of singlet oxygen.

Figure 6.
Singlet oxygen generation from H_2O_2 by NaOCl.

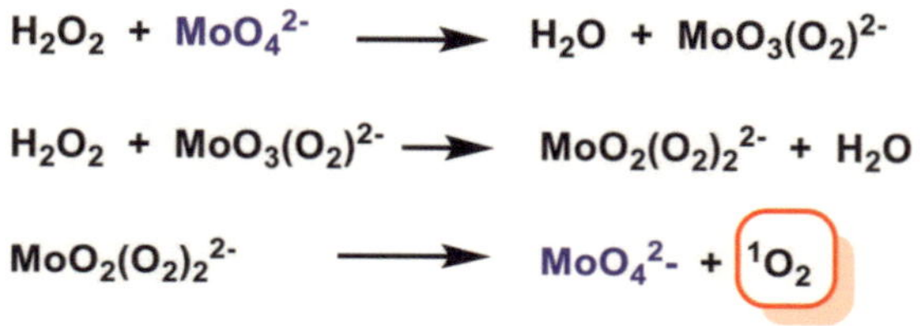

Figure 7.
Singlet oxygen generation from H_2O_2 by the $[MoO_4]^{2-}$ catalyst.

3.1 Chemical generation

There are many chemical reactions, one of the best-known examples being the reaction of NaClO with H_2O_2. The decomposition reaction of H_2O_2 in the presence of sodium hypochlorite or sodium hypobromite is the oldest reaction [9] for the formation of singlet oxygen, **Figure 6**.

$^1O_2$ is formed from the chlorine peroxide anion ($ClOO^-$), which is an intermediate in the above reaction. The best results are achieved when H_2O_2 decomposition is performed using a MoO_4^{2-} catalyst [10–13], **Figure 7**.

Among the inorganic peroxides, calcium peroxide diperoxohydrate (CaO_2-$2H_2O_2$), which is easily prepared by treating $CaCl_2$ or $Ca(OH)_2$ with H_2O_2 is of special interest for its ability to release $^1O_2$. In the first step, $Ca(OH)_2$ is converted into CaO_2-$8H_2O$, which takes up two H_2O_2 molecules giving CaO_2-$2H_2O_2$. By thermolysis at moderate temperature, the latter peroxide decomposes into $^1O_2$ and CaO_2-$8H_2O$. This catalytic cycle would operate until complete disproportionation of H_2O_2 and leave CaO_2-$8H_2O$ as the final product [14], **Figure 8**.

Singlet oxygen can also be generated chemically by the decomposition of ozonides of triphenylphosphite [15], **Figure 9**.

$^1O_2$ production from acyl peroxides such as benzoyl peroxide and lauryl peroxide [16, 17], **Figure 10**.

By decomposition of aromatic endoperoxides, singlet oxygen is also generated chemically. Certain dienes (anthracenes, naphthalenes, and pyridones) react with

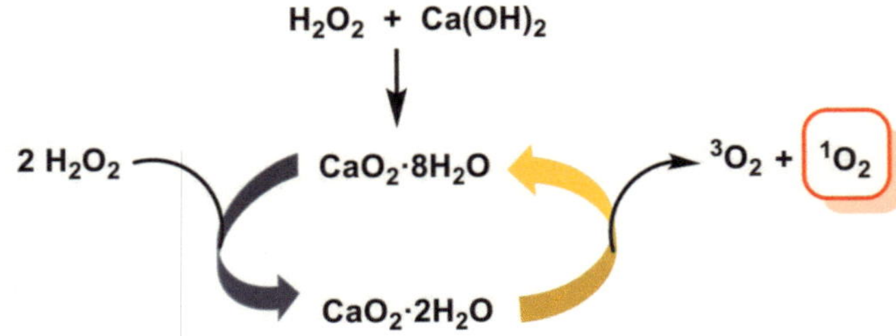

Figure 8.
Singlet oxygen generation from CaO_2-$2H_2O_2$.

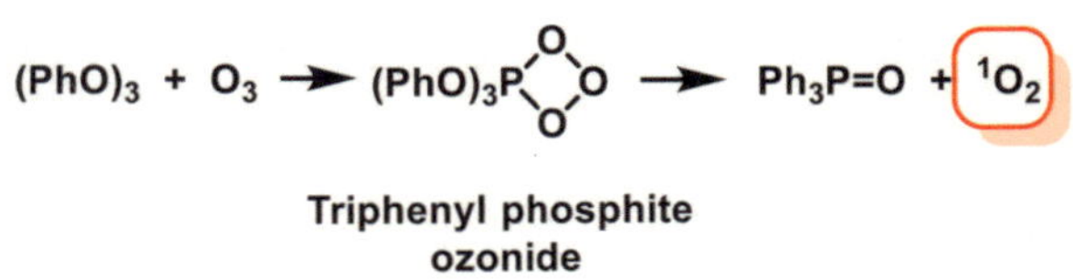

Figure 9.
Singlet oxygen generation from ozonides of tri-phenyl-phosphites.

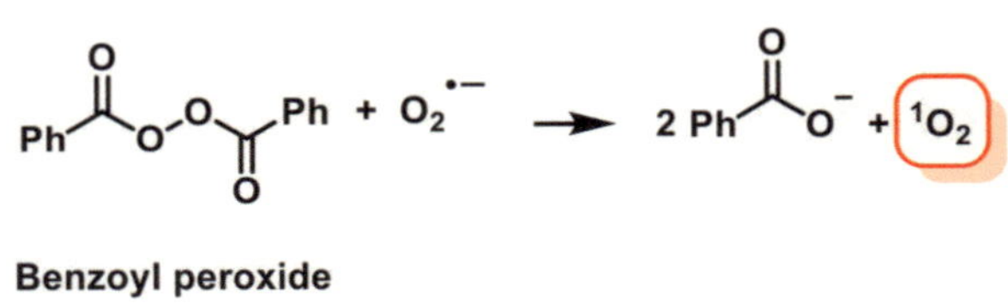

Figure 10.
Singlet oxygen generation from benzoyl peroxide.

$O_2(^3\Sigma_g\text{-})$ *via* sensitized photooxygenation to produce endoperoxides, which by heating dissociate and regenerate the starting hydrocarbon and $^1O_2$. The thermal stability of endoperoxides depends on the structure of the aromatic or heteroaromatic hydrocarbon. These organic endoperoxides are called reversible molecular singlet oxygen transporters. The lifetime of the endoperoxide precursors depends on the temperature and chemical structure of the dienes. Derivatives of naphthalenes, anthracenes, and pyridones are among the most studied examples that can undergo reversible endoperoxide formation for this purpose, **Figure 11**. All carriers are based on the reversible [4 + 2] cycloaddition of $^1O_2$ to a diene and are therefore generated by photooxygenation.

3.2 Photochemical generation

The absorption of light by a photosensitizer in its ground state (^{1}Sens) transforms it into an excited singlet state (^{1}Sens*). This short-lived state is rapidly converted through intersystem crossover into a more stable and longer-lived species, the excited triplet state (^{3}Sens*). The photosensitizer in its triple-excited state has the ability to transfer its energy to the $^3O_2$ dissolved in the medium. As a consequence of this transfer, the sensitizer is regenerated in its basal state, and the O_2 remains in its singlet excited state, **Figure 12**.

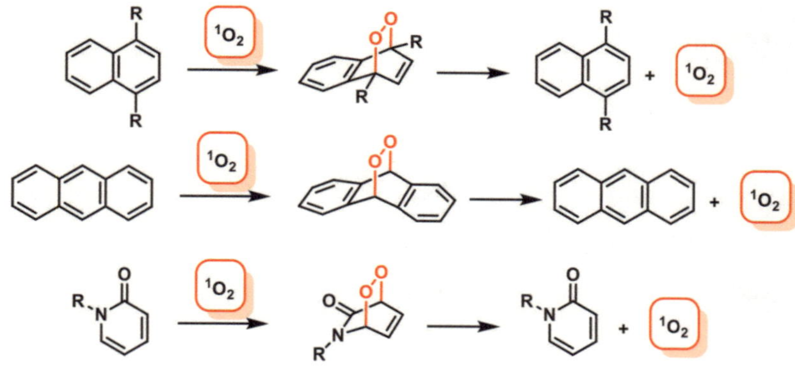

Figure 11.
Generation of singlet oxygen based on the formation and decomposition of endoperoxides.

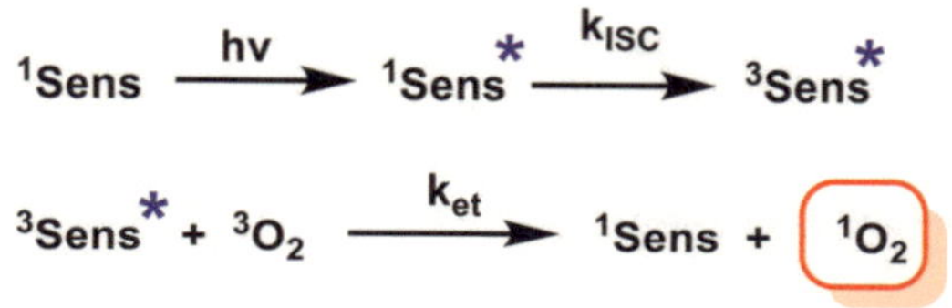

Figure 12.
The most widely used method in the generation of $O_2(^1\Delta_g)$ is photosensitization.

Photosensitization represents the most convenient, safe, easy, and environmentally friendly method for $^1O_2$ generation. The generation of singlet oxygen through photosensitization has been widely exploited in photodynamic therapy, environmental remediation, and synthesis [18].

3.3 Possible production of singlet oxygen in vivo

Some of the possible biological sources of $^1O_2$ generation are shown in **Figure 13**.

Possible biological sources of $O_2(^1\Delta_g)$ include (i) reactions catalyzed by peroxidases (myeloperoxidase) or oxygenases (lipoxygenases) [19, 20], (ii) recombination of peroxyl radicals can lead to the release of $^1O_2$ as a result of the decomposition of transient tetroxides according to Russell's concerted mechanism (Russell reaction) [6, 21], (iii) oxidation with ozone of amino acids, peptides, and proteins, (iv) hydrogen peroxide reactions with hypochlorite or peroxynitrite, (v) ozone oxidation of amino acids, peptides, and proteins [22], (vi) hydrogen peroxide reactions with hypochlorite or peroxynitrite [23, 24], (vii) thermolysis of endoperoxides [25–35], (vii) *in vitro* photodynamic processes involving type-II photosensitization reactions using suitable dyes [36–38], and (viii) UV irradiation of aromatic amino acids in proteins and immunoglobulins.

4. Applications

$^1O_2$ has been gaining much attention due to its pivotal role in a wide variety of chemical and biological processes, for example, plant signaling, organic synthesis, oxidation of food and beverages, or photodynamic therapy [39, 40], **Figure 14**.

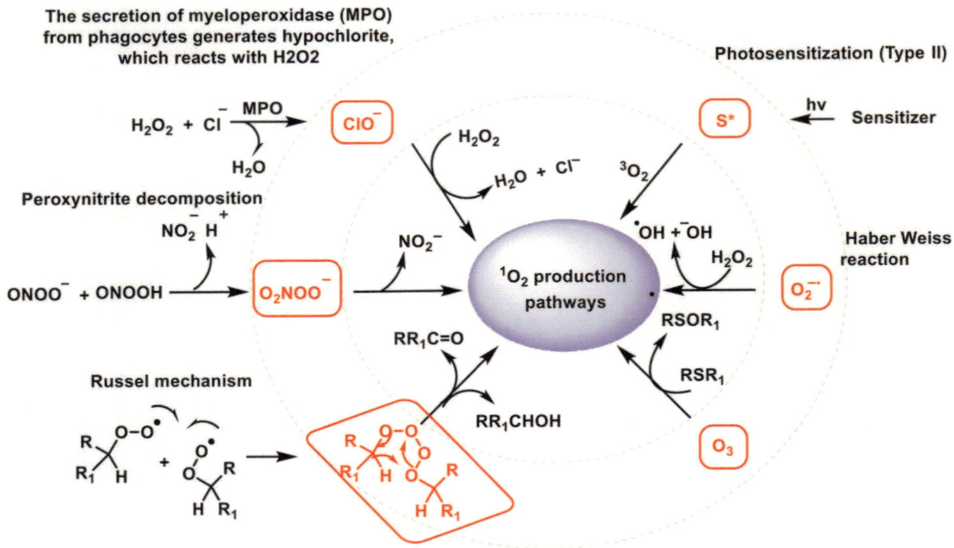

Figure 13.
Generation of singlet oxygen in biological systems.

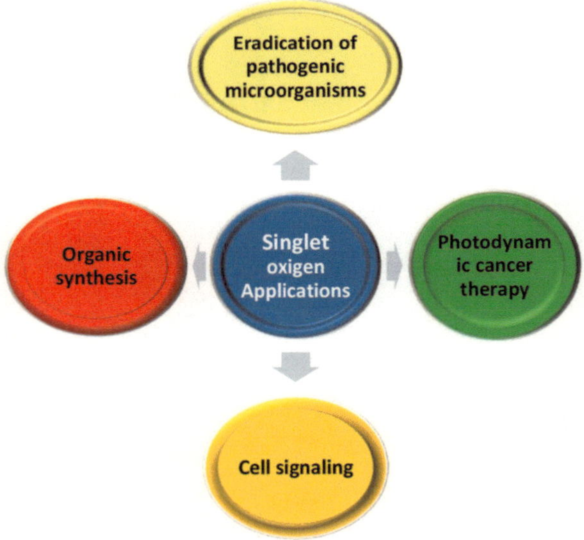

Figure 14.
The following sections outline different applications that have been developed over the last decades.

4.1 Photodynamic therapy (PDT)

Photodynamic therapy (PDT) has received increasing attention for the treatment of cancer and other diseased tissues as this methodology allows noninvasive, selective, and localized destruction of tumor cells with reduced side effects. PDT consists of the administration of a photosensitizer, which is selectively accumulated in certain cells or

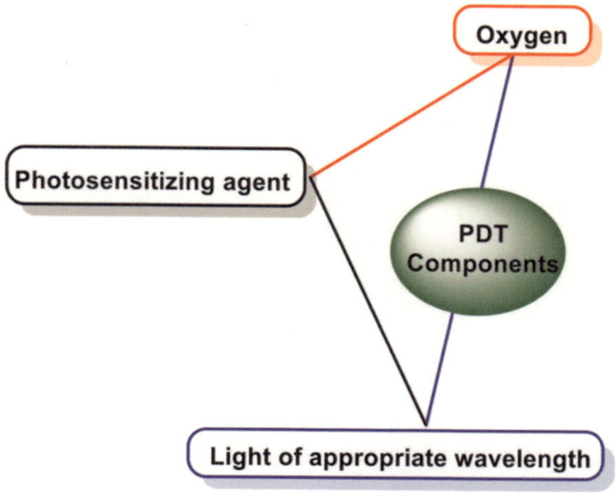

Figure 15.
PDT components.

tissues so that with subsequent irradiation and in the presence of oxygen, it triggers photooxidation of biological materials and subsequent cell death.

PDT is composed of three components, **Figure 15**. Each of these components alone lacks curative properties. The contribution of each depends on the type and dose of PS, the time between administration and exposure to light, the intensity of light, and the concentration of oxygen generated.

PDT combines light, molecular oxygen, and a photosensitizer (PS) for the production of reactive oxygen species (ROS) such as singlet oxygen ($^1O_2$) and free radicals that induce oxidative stress and eventually cell death. PDT was the first example of a drug-device combination approved by the Food and Drug Administration (FDA) for the production of reactive oxygen species (ROS) such as singlet oxygen ($^1O_2$) and free radicals that induce oxidative stress and eventually cell death [41]. PDT was the first example of a drug-device combination approved by the Food and Drug Administration (FDA) [42].

4.2 Light sources

Most Ps are activated with red light between 630 and 700 nm, which has a tissue penetrating power between 0.5 cm (at 630 nm) and 1.5 cm (at 700 nm) [43]. Lasers and light-emitting diodes (LEDs) coupled to flexible fiber optic devices are used as light sources. Depending on the depth of the pathologies to be treated, LEDs can be implanted at the end of different catheters to form screens for irradiation of large areas. The wavelength required will be determined by the PS with which it is to be associated and applied in the most focal way possible on the tissue to be treated. However, the most commonly used light sources in photodynamic treatments are lasers.

4.3 Photosensitizing agent in photodynamic therapy

A PS is defined as a compound capable of absorbing light and subsequently triggering a photophysical or photochemical reaction in response to it. For clinical use, it must meet a number of characteristics as summarized in **Figure 16** [39, 44].

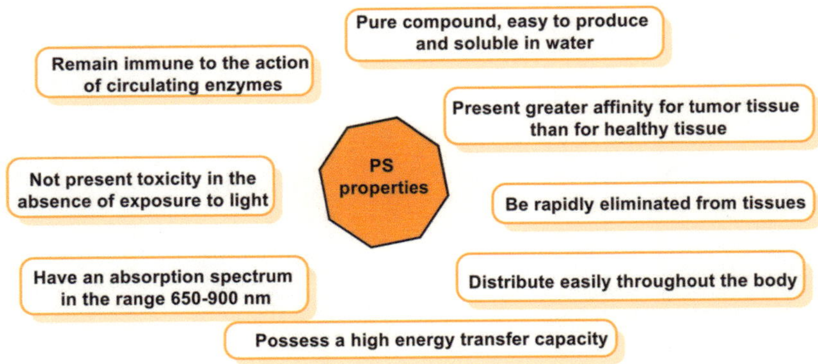

Figure 16.
PS properties.

The PS most suitable for PDT can be divided into two main groups: porphyrinoid or non-porphyrinoid derivatives [44–47], **Figure 17**. Porphyrinoid-based derivatives are the most widely used in PDT applications as their extended π-systems confer unique photochemical characteristics that are valuable for their use as photosensitizers [48]. Within the porphyrinoid PS, they are usually classified into first-, second-, and third-generation photosensitizers depending on their evolution [49]. Natural porphyrins and their derivatives constitute the first generation of PS. Most of the PS under investigation for the treatment of cancer and other diseases is based on the tetrapyrrole core that includes porphyrins, chlorins, bacterio-chlorins, phthalocyanines, and texaphyrins. These molecules have been chosen for their low toxicity in the absence of light to mammalian and animal cells and for their tumor-localizing properties. PS that have been studied for their ability to kill microorganisms are halogenated xanthenes such as rose bengal (RB), chlorinated poly-L-lysine-chlorinacorinate, and phenothiazines such as toluidine blue O (TBO) and methylene blue [50] and poly-L-lysine-chlorinate conjugates [51].

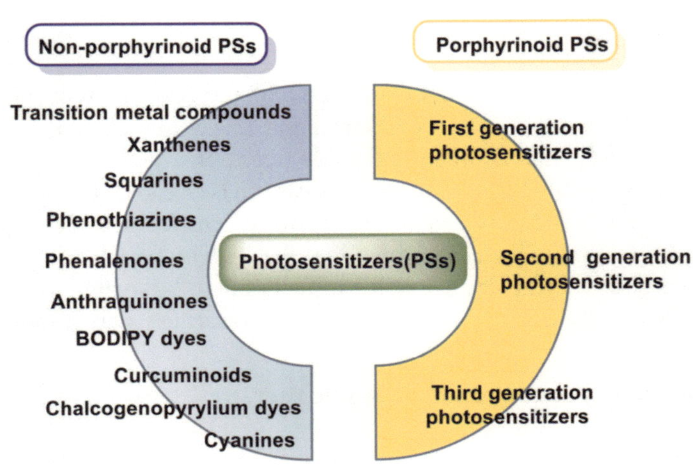

Figure 17.
Types of photosensitizers.

4.4 First-generation photosensitizers

First-generation PSs are a complex mixture of compounds among which hemato-porphyrin (HpD) and photofrin (R) derivatives are the representatives of the first-generation Fs. Photofrin is one of the few photosensitizers approved for the treatment of early and advanced esophageal and lung cancer by the US FDA and by various health agencies worldwide. It is now being extended to the therapy of different oncological pathologies of the head, neck, abdomen, thorax, brain, bowel, cervical, skin, and breast Photofrin® (as HPD) causes persistent photosensitization of the skin. This required avoidance of intense sunlight for at least 30 days after drug administration. HpDs are a complex mixture of monomers, dimers, and oligomers, linked together by the formation of ethers, esters, and carbon-carbon bonds [44], **Figure 18**.

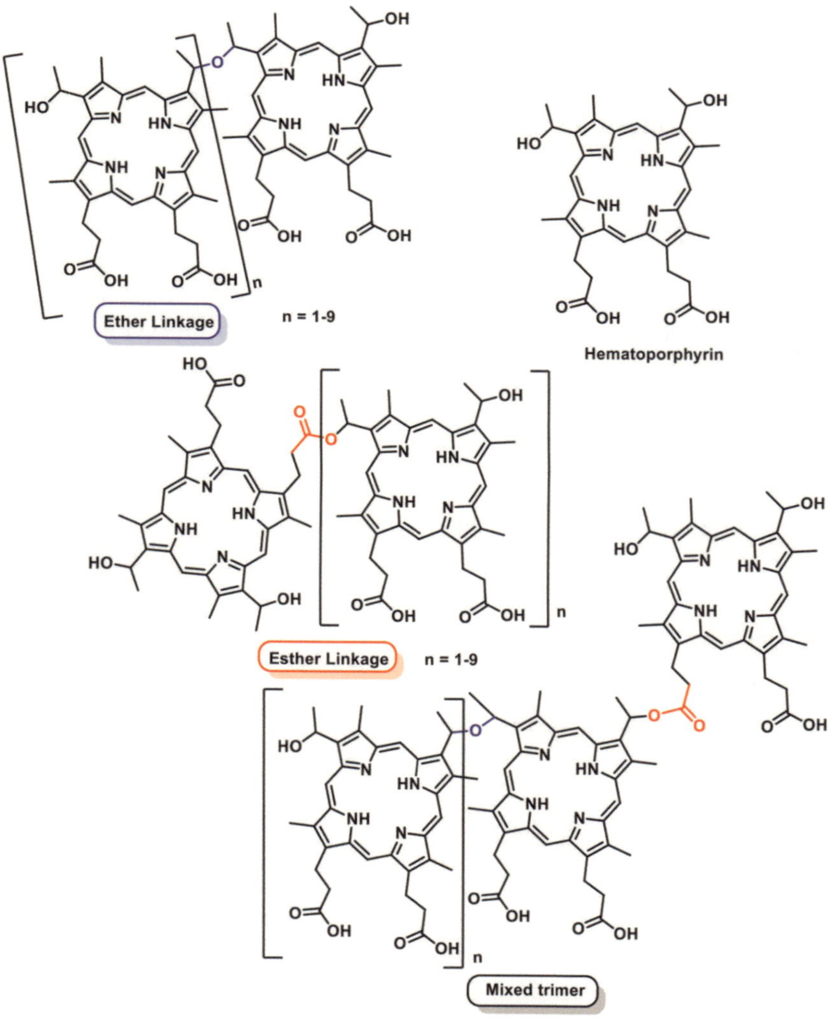

Figure 18.
First-generation porphyrin derivatives.

4.5 Second-generation photosensitizers

Second-generation photosensitizers have been developed since the late 1980s. They are pure compounds of known chemical structure with absorption maxima at wavelengths above 630 nm, high molar extinction coefficients, quantitative $^1O_2$ formation, and a reduction in the side effects and undesirable properties of the first generation. The major disadvantage of these is that they are highly lipophilic, which hinders their bioavailability, favors self-aggregation, and poses a challenge for the development of a pharmaceutical form for intravenous administration.

4.5.1 Porphyrin photosensitizers

Porphyrin derivatives such as meta-tetra-(hydroxyphenyl)-porphyrin (m-THPP, Foscan, or Temoporphyrin) and 5,10,15,20-tetrakis(4-sulfonatophenyl)-21H-23H-porphyrin (TPPS4). These compounds are pure, extremely potent structures, which are photoactivated at wavelengths longer than those corresponding to HpD, with higher molar extinction coefficients. This group also includes macrocyclic porphyrin derivatives such as chlorines, bacterio-chlorines, phthalocyanines, porphyrins, and texaphyrins. Due to the potency of the second-generation Fs, the drug doses and light intensity required to obtain a photofrin(R)-like response are up to 100 times lower, **Figure 19**.

4.5.2 Non-porphyrin photosensitizers

The development of non-porphyrin PSs, **Figure 20**, for application in photodynamic therapy lags considerably behind the evolution of porphyrin derivatives. Cationic compounds, such as phenothiazines, have high molar extinction

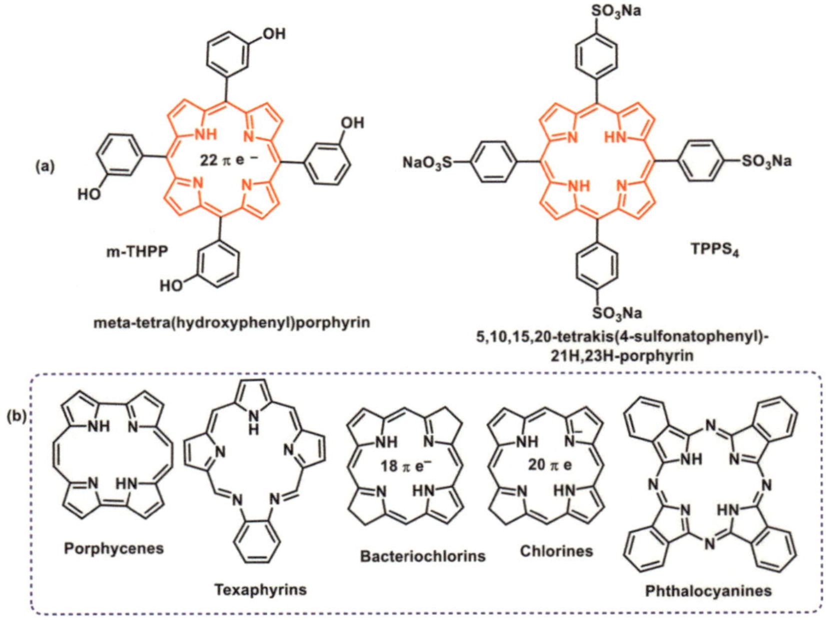

Figure 19.
(a) Second-generation porphyrin derivatives and (b) Second-generation porphyrin macrocycles.

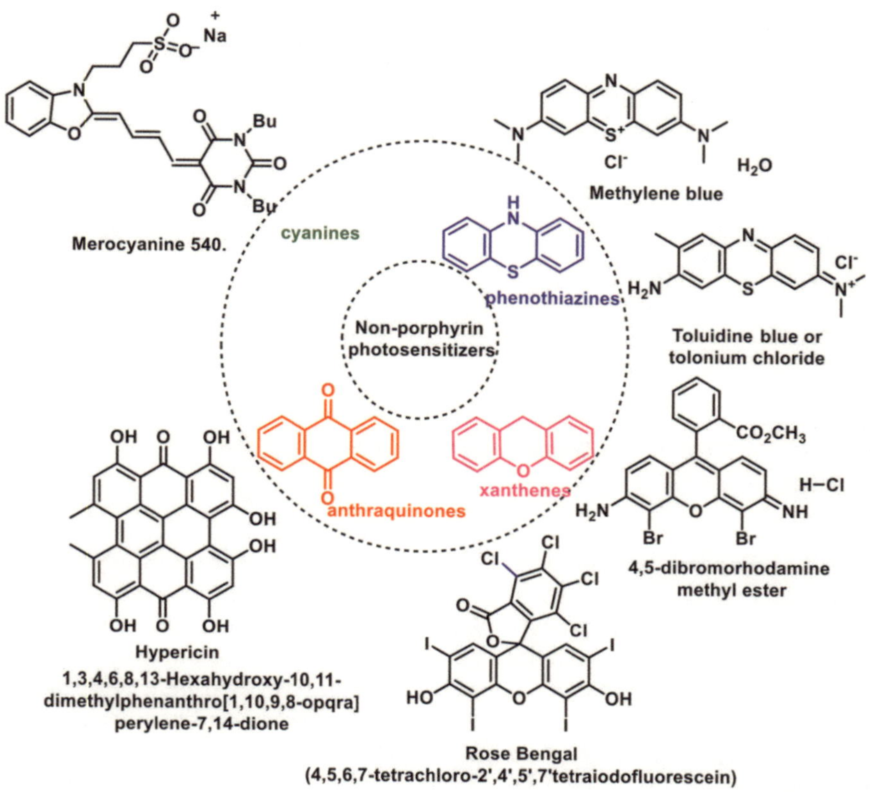

Figure 20.
Chemical structure of non-porphyrin photosensitizers.

coefficients between 600 and 800 nm, including methylene blue and toluidine blue. Methylene blue has been applied in the clinical treatment of basal cell carcinoma and Kaposi's sarcoma. In addition, there are *in vitro* assays of adenocarcinoma, bladder carcinoma, and cervical tumor cells. Toluidine blue has been extensively studied for the inactivation of various pathogenic microorganisms. Both thiazine derivatives are currently being evaluated for the treatment of chronic periodontitis. Literature has shown that the mitochondrial membrane potential of tumor cells is more negative than that of normal cells so that cationic PSs are retained mainly in the cells to be treated. The bacterial cell wall has anionic components that readily interact with cationic drugs.

Rose bengal has shown good results in the photodynamic treatment of breast carcinoma and metastatic melanoma. 4,5-dibromorhodamine methyl ester is effective against graft-versus-host disease, resulting in the destruction of lymphocytes by apoptosis.

Hypericin is a natural anthraquinone with defined photochemical properties and shows selectivity for tumor cells. Merocyanin 540 is the representative of the cyanins used in the treatment of leukemia and neuroblastoma.

4.5.3 Endogenous photosensitizer: precursors

Protoporphyrin IX (PpIX) is an endogenous photosensitizer. Under normal conditions, PpIX is present at an extremely low concentration to produce a photosensitizing

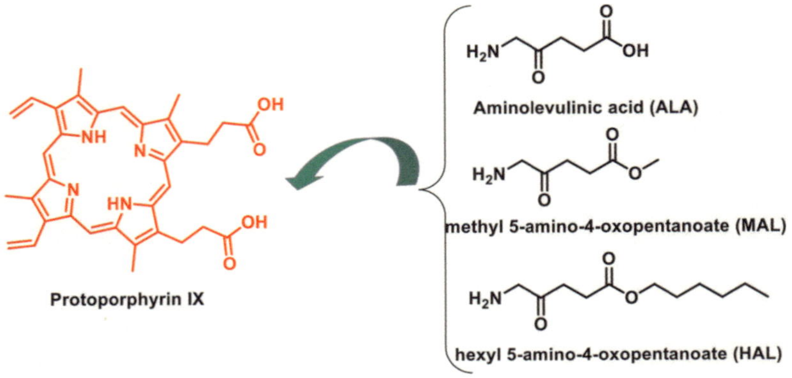

Figure 21.
Chemical structure of Protoporphyrin IX and of ALA MAL and HAL.

reaction. To stimulate the synthesis of protoporphyrin IX, it is necessary to administer an excess of amino levulinic acid (ALA) and its derivatives, ALA-methyl ester (MAL) and ALA hexyl ester (HAL), thus stimulating the synthesis of PpIX, which can accumulate in damaged tissue and be applied in the photodynamic treatment of different pathologies. It has been used in the treatment of actinic keratosis, carcinoma, basal cell carcinoma, Bowen's disease, and bladder cancer. PDT with ALA and its derivatives, ALA-methyl ester (MAL) and ALAhexyl ester (HAL), is an approved treatment for a number of malignant and premalignant conditions (**Figure 21**).

4.6 Third-generation photosensitizers

Third-generation photosensitizers are called nano photosensitizers and are made up of a combination of PSs and different vector systems. They are proposed as a solution to overcome the problems of low water solubility of the first- and second-generation Ps as most of them have an aromatic character.

Numerous delivery strategies have been evaluated with the aim of achieving a highly selective and effective therapy. Among the drug carrier systems used in the development of third-generation PSs, applicable in PDT are polymeric nanoparticles, solid lipid nanoparticles, nanostructured lipid carriers, gold nanoparticles, hydrogels, liposomes, liquid crystals, dendrimers, and cyclodextrin.

4.6.1 Liposomes

One of the most recent and successful applications of liposomes is to serve as delivery vehicles for photodynamic agents to improve their solubility, specificity, bioavailability, and tendency to aggregate. Therefore, liposomal photosensitizers as third-generation formulations have shown great potential to increase the efficacy of photodynamic cancer therapy and consist of concentric vesicles formed by one or more concentric natural or synthetic phospholipid bilayers enclosing an aqueous compartment. Nontoxic, highly biocompatible, and biodegradable, these delivery systems have allowed increasing the activity of Ps *in vitro* and *in vivo*, reducing their toxicity and improving their biodistribution [52].

4.6.2 Micelles

Micelles are monolayer structures formed by amphiphilic surfactants, also called amphipathic surfactants, which are molecules that have one hydrophilic end, that is they are soluble in water, and one hydrophobic end, which means that they repel water.

Polymeric micelles have been widely used in recent years due to their higher stability compared to micellar systems based on conventional surfactants. They are prepared by association of copolymers dispersed in an aqueous medium, forming particles with diameters of less than 100 nm [53]. Polymeric micelles are presented as an alternative for the delivery of hydrophobic Ps and have the following advantages (i) simple preparation, (ii) efficient drug loading without chemical modification, (iii) controlled release [54], and (iv) no side effects of skin photosensitivity [55].

4.6.3 Solid nanoparticles

Solid nanoparticles are a promising new tool for drug delivery in PDT [56, 57]. The advantages over second-generation Ps can be summarized as (i) ability to deliver a large amount of drug to target cells, (ii) prevent degradation in the biological environment, (iii) incorporate multiple components as contrast agents, and (iv) incorporate of different ligands in order to increase selectivity [58].

The solid nanoparticles, most commonly used in the development of third-generation Ps, are polymeric, silica, and gold nanoparticles [57, 59]. Silica nanoparticles have numerous advantages over polymeric systems such as their nontoxic nature, inert, and stable properties. As they are nonbiodegradable, particle size control is necessary to ensure their elimination from the body *via* the kidney and to avoid their uptake by the reticuloendothelial system. Due to their high porosity, silica nanoparticles have a high PS loading capacity and contribute significantly to the efficacy of PDT.

Gold nanoparticles increase the field of incident light around them, which could increase the excitation efficiency of the PS they carry [60]. The vehiculization of porphyrins, phthalocyanines, and thiazines on gold nanoparticles increased (i) blood circulation time, (ii) selective delivery to tumor tissues, (iii) $^1O_2$ generation yield, and (iv) photodynamic activity of free PSs [61].

5. Mechanism of action of photosensitizers

Most PS in their ground state has two electrons with opposite spins. Light absorption leads to the transfer of an electron to a higher energy orbital. This excited PS is very unstable and emits this excess energy in the form of fluorescence and/or heat. Another possibility is that an excited PS can undergo cross-system crossover to form a triplet state. The photosensitizer in the triplet state can decay without radiation to the ground state or transfer its energy to molecular oxygen. This step leads to the formation of singlet oxygen, and the reaction is called a type-II process [62]. A type-I process can also occur [63], **Figure 22**.

In the type-I process, electron transfer occurs between the activated PS and the surrounding molecules leading to the release of free radicals. These radicals are highly active and interact with oxygen-producing endogenous molecules to produce anion

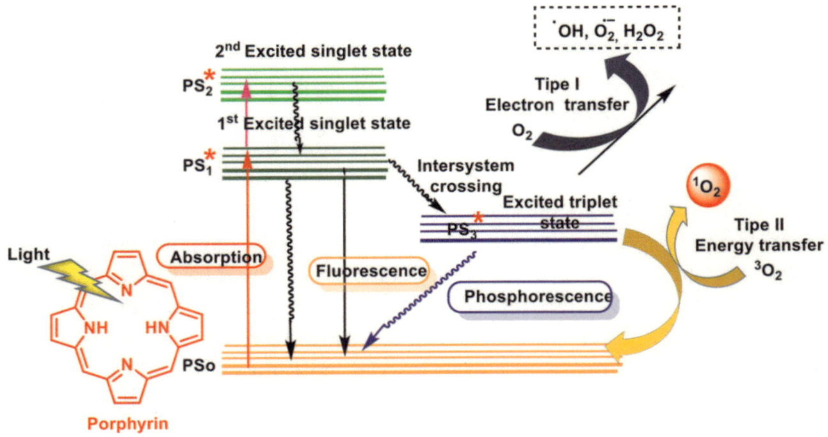

Figure 22.
Dynamic singlet oxygen therapy: Sensitizer diagram.

superoxide, hydroxyl radical, hydrogen peroxide, and free radicals. These ROS cause damage to the integrity of cell membranes and their internal structures.

The type-II process, this is the interaction of PS with oxygen. For this reaction to take place, PS must be in its triplet form, activating oxygen to its active or singlet form ($^1O_2$), which allows it to interact with a large number of substances directly such as amino acids or lipids. The short half-life of this singlet oxygen (<0.4 milliseconds) means that its diffusion range is limited to 45 nm in the cellular medium, and destruction is only limited to the intracellular structures, it can access [64, 65].

6. Clinical applications of the PDT

PDT is a minimally invasive outpatient therapeutic modality that has low toxicity and can be applied repeatedly at the site of action. In oncology, it can be combined with chemotherapy, ionizing radiation, or surgery. PDT is an approved technique for the treatment of some types of cancer and for antimicrobial therapies.

6.1 Photodynamic therapy of cancer

Photosensitizers (PS) can be used in various types of tumors. PS has been used as therapeutic agent for more than a century. The clinical use of eosin in the treatment of skin cancer can be given as an example of the first applications of PDT. The first objective was to treat skin tumors by using topical eosin [66, 67].

According to the World Health Organization (WHO), cancer is a group of cellular diseases characterized by unregulated cell growth. In the early stages, it is localized in healthy tissues and spreads to neighboring tissues or even to other organs of the body through metastatic cells. The idea of using photodynamic therapy (PDT) as a new treatment strategy was suggested in the early twentieth century. PDT has fewer side effects and toxicity than chemotherapy and/or radiotherapy. PDT is a particularly attractive alternative to conventional antitumor drugs due to its fundamental specificity and selectivity. This excitation causes the photosensitizer to generate singlet oxygen and other reactive oxygen species. PDT has been used in several types of

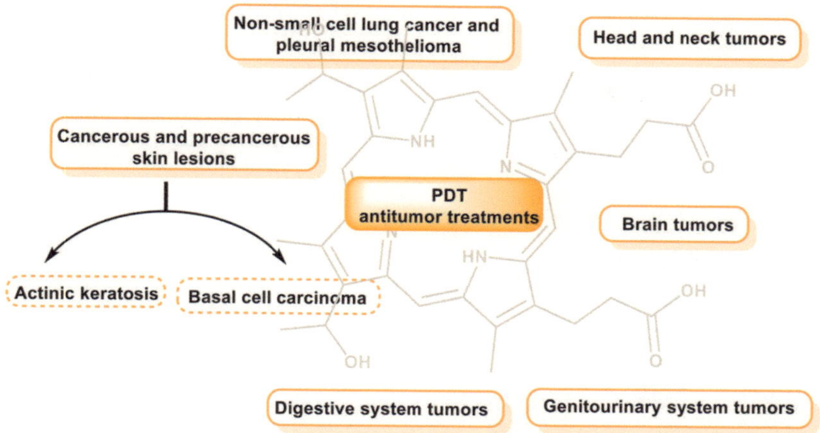

Figure 23.
PDT antitumor treatments.

cancer, including non-melanoma skin cancer, bladder cancer, esophageal cancer, head and neck cancer, and non-small cell lung cancer (NSCLC), **Figure 23**.

The PDT clinical procedure involves four stages, as shown in **Figure 24**.

1. Ps is administered to the patient, usually topically or intravenously.

2. Between 3- and 96-hours elapse before the affected area is illuminated. During this time, the drug is distributed throughout the body and selectively localized in the affected cells.

3. The affected area is irradiated locally using an appropriate irradiation source for photodynamic treatment.

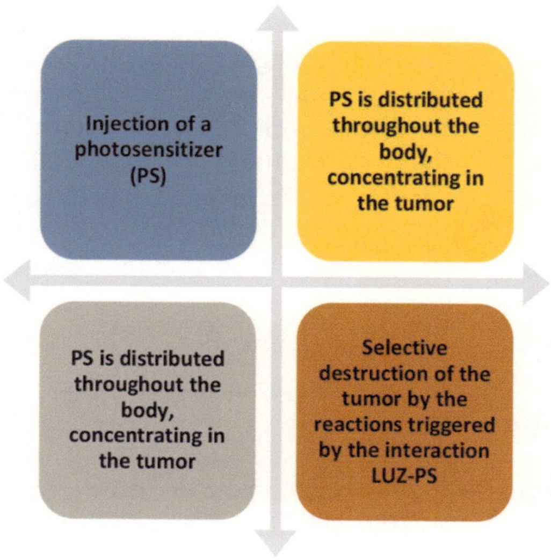

Figure 24.
Representation of an antitumor treatment with PDT.

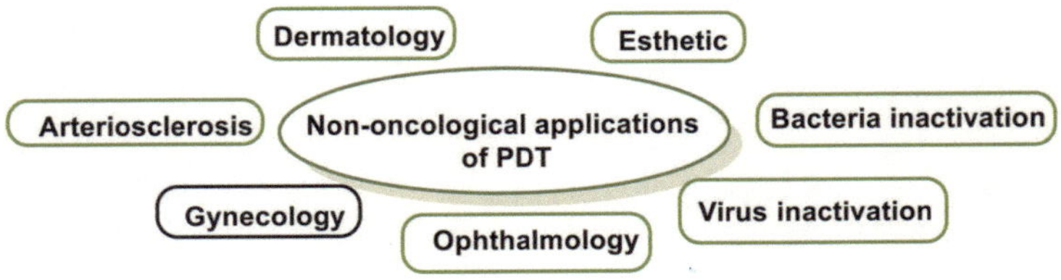

Figure 25.
Non-oncological applications of PDT.

4. A series of intracellular reactions are triggered, which give rise to the formation of ROS that produce irreversible biological damage in the treated area, leading to cell death. Cell destruction can be triggered by different pathways [44, 68].

6.2 Non-oncological applications of PDT

PDT has been mainly focused on cancer treatment, but in recent years, new applications in different fields are being evaluated, **Figure 25**.

Esthetics: Photo depilation is based on using a laser that irradiates at an appropriate wavelength (depending on the color of the hair and the color of the skin) to permanently remove the hair. The action of topically applied ALA is currently being studied to improve the photo-depilation technique.

Dermatology: PDT is used to treat non-oncological dermatological diseases such as psoriasis, vascular malformations, and acne [69–71].

Inactivation of bacteria: The increasing resistance of bacteria to antibiotics has led to the development of alternative antimicrobial techniques. Certain bacteria can be inactivated by being illuminated after an incubation period with certain porphyrins and phthalocyanines. This new application may be of great use in the treatment of internal cavity cleaning and in the treatment of oral conditions.

Virus inactivation: Photoinactivation of viruses in human blood as a method to sterilize blood and blood products for transfusion [72].

Ophthalmology: Age-related macular degeneration (AMD) is a leading cause of blindness and is due to the rapid abnormal growth of blood vessels in the retina. The results obtained by treating this disease with PDT show a hopeful future.

Arteriosclerosis: The possibility of treating this disease with PDT is based on the fact that atheroma plaques in damaged arteries retain higher concentrations of porphyrins than the normal vascular wall [69].

Gynecology: PDT is an alternative to hysterectomy for women with dysfunctional uterine bleeding [73, 74].

Funding information

This research was funded by Agencia Canaria de Investigación, Innovación y Sociedad de la Información (ACIISI) del Gobierno de Canarias, Project ProID2020010134, Caja Canarias, Project 2019SP43, the Spanish Ministry of Economy and Competitiveness (Grant PID2019-105838RB-C31) and the State Plan for Scientific, Technical Research and Innovation 2021–2023 from the Spanish Ministry of

Science and Innovation (project PLEC2022-009507) and the State Plan for Scientific, Technical Research and Innovation 2021–2023 from the Spanish Ministry of Science and Innovation (project PLEC2022-009507).

Author details

Celia María Curieses Andrés[1], José Manuel Pérez de la Lastra[2]*, Celia Andrés Juan[3], Francisco J. Plou[4] and Eduardo Pérez-Lebeña[5]

1 Hospital Clínico Universitario, Avenida de Ramón y Cajal, Valladolid, Spain

2 Institute of Natural Products and Agrobiology, CSIC-Spanish Research Council, La Laguna, Spain

3 Cinquima Institute and Department of Organic Chemistry, Faculty of Sciences, Valladolid University, Valladolid, Spain

4 Institute of Catalysis and Petrochemistry, CSIC-Spanish Research Council, Madrid, Spain

5 Sistemas de Biotecnología y Recursos Naturales, Valladolid, Spain

*Address all correspondence to: jm.perezdelalastra@csic.es

References

[1] Wilkinson F, Helman WP, Ross AB. Rate constants for the decay and reactions of the lowest electronically excited singlet state of molecular oxygen in solution. An expanded and revised compilation. Journal of Physical and Chemical Reference Data. 1995;**24**:663-677. DOI: 10.1063/1.555965

[2] Schweitzer C, Schmidt R. Physical mechanisms of generation and deactivation of singlet oxygen. Chemical Reviews. 2003;**103**:1685-1758. DOI: 10.1021/cr010371d

[3] Volman DH, Hammond GS, Gollnick K. Advances in Photochemistry. Hoboken, NJ, USA: John Wiley & Sons Inc.; 2009

[4] Frimer AA. The reaction of singlet oxygen with olefins: the question of mechanism. Chemical Reviews. 1979;**79**: 359-387

[5] Cadet J, Di Mascio P. Peroxides in biological systems. Patai's Chemistry of Functional Groups. 2009. DOI: 10.1002/ 9780470682531.pat0357

[6] Miyamoto S, Martinez GR, Medeiros MH, Di Mascio P. Singlet molecular oxygen generated from lipid hydroperoxides by the Russell mechanism: studies using 18O-labeled linoleic acid hydroperoxide and monomol light emission measurements. Journal of the American Chemical Society. 2003;**125**:6172-6179

[7] Miyamoto S, Nantes IL, Faria PA, Cunha D, Ronsein GE, Medeiros MH, et al. Cytochrome c-promoted cardiolipin oxidation generates singlet molecular oxygen. Photochemical & Photobiological Sciences. 2012;**11**: 1536-1546

[8] Miyamoto S, Martinez GR, Rettori D, Augusto O, Medeiros MH, Di Mascio P. Linoleic acid hydroperoxide reacts with hypochlorous acid, generating peroxyl radical intermediates and singlet molecular oxygen. Proceedings of the National Academy of Sciences. 2006; **103**:293-298

[9] Murray RW. Chemical Sources of Singlet Oxygen. 1979

[10] Aubry J. Search for singlet oxygen in the decomposition of hydrogen peroxide by mineral compounds in aqueous solutions. Journal of the American Chemical Society. 1985;**107**:5844-5849

[11] Aubry J, Cazin B. Chemical sources of singlet oxygen. 2. Quantitative generation of singlet oxygen from hydrogen peroxide disproportionation catalyzed by molybdate ions. Inorganic Chemistry. 1988;**27**:2013-2014

[12] Niu Q, Foote C. Singlet molecular oxygen generation from the decomposition of sodium peroxotungstate and sodium peroxomolybdate. Inorganic Chemistry. 1992;**31**:3472-3476

[13] Nardello V, Marko J, Vermeersch G, Aubry J. 90Mo NMR and kinetic studies of peroxomolybdic intermediates involved in the catalytic disproportionation of hydrogen peroxide by molybdate ions. Inorganic Chemistry. 1995;**34**:4950-4957

[14] Nardello V, Aubry J-M, Briviba K, Sies H. Identification of the precursor of singlet oxygen (1 O 2, 1 Δ g) involved in the disproportionation of hydrogen peroxide catalyzed by calcium hydroxide. Chemical Communications. 1998:599-600. DOI: 10.1039/A708161H

[15] CaMinade AM, Khatib FE, Koenig M, Aubry JM. Ozonides de phosphite source d'oxygène singulet: rendement, mécanisme. Canadian Journal of Chemistry. 1985;**63**:3203-3209

[16] Danen WC, Arudi RL. Generation of singlet oxygen in the reaction of superoxide anion radical with diacyl peroxides. Journal of the American Chemical Society. 1978;**100**:3944-3945

[17] Khan AU, Kasha M. Singlet molecular oxygen in the Haber-Weiss reaction. Proceedings of the National Academy of Sciences. 1994;**91**: 12365-12367

[18] DeRosa MC, Crutchley RJ. Photosensitized singlet oxygen and its applications. Coordination Chemistry Reviews. 2002;**233**:351-371

[19] Kanofsky J, Axelrod B. Singlet oxygen production by soybean lipoxygenase isozymes. Journal of Biological Chemistry. 1986;**261**: 1099-1104

[20] Klebanoff SJ. Myeloperoxidase: friend and foe. Journal of Leukocyte Biology. 2005;**77**:598-625

[21] Russell GA. Deuterium-isotope effects in the autoxidation of aralkyl hydrocarbons. mechanism of the interaction of peroxy radicals1. Journal of the American Chemical Society. 1957; **79**:3871-3877

[22] Kanofsky J, Sima P. Singlet oxygen production from the reactions of ozone with biological molecules. Journal of Biological Chemistry. 1991;**266**: 9039-9042

[23] Di Mascio P, Bechara EJ, Medeiros MH, Briviba K, Sies H. Singlet molecular oxygen production in the reaction of peroxynitrite with hydrogen peroxide. FEBS Letters. 1994;**355**:287-289

[24] Martinez GR, Di Mascio P, Bonini MG, Augusto O, Briviba K, Sies H, et al. Peroxynitrite does not decompose to singlet oxygen (1ΔgO2) and nitroxyl (NO−). Proceedings of the National Academy of Sciences. 2000;**97**: 10307-10312

[25] Moureu C, Dufraisse C, Dean P. An organic peroxide separable from the rubrene peroxide. Comptes Rendus Hebdomadaires des Séances de l'Académie des Sciences. 1926;**182**: 1584-1587

[26] Dufraisse C, Velluz L. The labile union of carbon and peroxide which spontaneously dissociates in the cold. Comptes Rendus Hebdomadaires des Séances de l'Académie des Sciences. 1939;**208**:1822-1824

[27] Mano CM, Prado FM, Massari J, Ronsein GE, Martinez GR, Miyamoto S, et al. Excited singlet molecular O2 (1Δg) is generated enzymatically from excited carbonyls in the dark. Scientific Reports. 2014;**4**:5938

[28] Wasserman HH, Larsen DL. Formation of 1, 4-endoperoxides from the dye-sensitized photo-oxygenation of alkyl-naphthalenes. Journal of the Chemical Society, Chemical Communications. 1972:253-254. DOI: 10.1039/C39720000253

[29] Di Mascio P, Sies H. Quantification of singlet oxygen generated by thermolysis of 3, 3′-(1, 4-naphthylene) dipropionate endoperoxide. Monomol and dimol photoemission and the effects of 1, 4-diazabicyclo [2.2.2] octane. Journal of the American Chemical Society. 1989;**111**:2909-2914

[30] Pierlot C, Aubry J-M, Briviba K, Sies H, Di Mascio P. [1] Naphthalene endoperoxides as generators of singlet oxygen in biological media. Methods in Enzymology. 2000;**319**:3-20

[31] Saito I, Matsuura T, Inoue K. Formation of superoxide ion from singlet oxygen. Use of a water-soluble singlet oxygen source. Journal of the American Chemical Society. 1981;**103**:188-190

[32] Ravanat J-L, Di Mascio P, Martinez GR, Medeiros MH, Cadet J. Singlet oxygen induces oxidation of cellular DNA. Journal of Biological Chemistry. 2000;**275**:40601-40604

[33] Ravanat J-L, Sauvaigo S, Caillat S, Martinez GR, Medeiros MHG, Mascio PD, Favier A, Cadet J. Singlet oxygen-mediated damage to cellular DNA determined by the comet assay associated with DNA repair enzymes. 2004;**385**(1):17-20. DOI: 10.1515/BC.2004.003

[34] Martinez GR, Ravanat J-L, Medeiros MH, Cadet J, Di Mascio P. Synthesis of a naphthalene endoperoxide as a source of 18O-labeled singlet oxygen for mechanistic studies. Journal of the American Chemical Society. 2000;**122**: 10212-10213

[35] Rolim JP, De-Melo MA, Guedes SF, Albuquerque-Filho FB, De Souza JR, Nogueira NA, et al. The antimicrobial activity of photodynamic therapy against Streptococcus mutans using different photosensitizers. Journal of Photochemistry and Photobiology B: Biology. 2012;**106**:40-46

[36] Yoon HY, Koo H, Choi KY, Lee SJ, Kim K, Kwon IC, et al. Tumor-targeting hyaluronic acid nanoparticles for photodynamic imaging and therapy. Biomaterials. 2012;**33**:3980-3989

[37] Bäumler W, Regensburger J, Knak A, Felgenträger A, Maisch T. UVA and endogenous photosensitizers–the detection of singlet oxygen by its luminescence. Photochemical & Photobiological Sciences. 2012;**11**:107-117

[38] BW, H. How does photodynamic therapy work? Photochemistry and Photobiology. 1991;**55**:145-157

[39] Agostinis P, Berg K, Cengel KA, Foster TH, Girotti AW, Gollnick SO, et al. Photodynamic therapy of cancer: an update. CA: A Cancer Journal for Clinicians. 2011;**61**:250-281. DOI: 10.3322/caac.20114

[40] Kim C, Meskauskiene R, Apel K, Laloi C. No single way to understand singlet oxygen signalling in plants. EMBO Reports. 2008;**9**:435-439

[41] Ding H, Yu H, Dong Y, Tian R, Huang G, Boothman DA, et al. Photoactivation switch from type II to type I reactions by electron-rich micelles for improved photodynamic therapy of cancer cells under hypoxia. Journal of Controlled Release. 2011;**156**:276-280

[42] Celli JP, Spring BQ, Rizvi I, Evans CL, Samkoe KS, Verma S, et al. Imaging and photodynamic therapy: Mechanisms, monitoring, and optimization. Chemical Reviews. 2010; **110**:2795-2838

[43] Veerendra NR, Rekha RK, Chandana G, Sangeeta S. Photodynamic therapy. Indian Journal of Dental Advancements. 2009;**1**:46-51

[44] Ormond AB, Freeman HS. Dye Sensitizers for Photodynamic Therapy. Materials (Basel). Mar 6 2013;**6**(3):817-840. DOI: 10.3390/ma6030817. PMID: 28809342; PMCID: PMC5512801

[45] Detty MR, Gibson SL, Wagner SJ. Current clinical and preclinical photosensitizers for use in photodynamic therapy. Journal of Medicinal Chemistry. 2004;**47**:3897-3915

[46] O'Connor AE, Gallagher WM, Byrne AT. Porphyrin and nonporphyrin

photosensitizers in oncology: preclinical and clinical advances in photodynamic therapy. Photochemistry and Photobiology. 2009;**85**:1053-1074

[47] Yano S, Hirohara S, Obata M, Hagiya Y, Ogura S-I, Ikeda A, et al. Current states and future views in photodynamic therapy. Journal of Photochemistry and Photobiology C: Photochemistry Reviews. 2011;**12**:46-67

[48] Dąbrowski JM, Pucelik B, Regiel-Futyra A, Brindell M, Mazuryk O, Kyzioł A, et al. Engineering of relevant photodynamic processes through structural modifications of metallotetrapyrrolic photosensitizers. Coordination Chemistry Reviews. 2016; **325**:67-101. DOI: 10.1016/j.ccr.2016.06.007

[49] Allison RR, Downie GH, Cuenca R, Hu XH, Childs CJ, Sibata CH. Photosensitizers in clinical PDT. Photodiagnosis and Photodynamic Therapy. 2004;**1**:27-42. DOI: 10.1016/s1572-1000(04)00007-9

[50] Usacheva MN, Teichert MC, Biel MA. Comparison of the methylene blue and toluidine blue photobactericidal efficacy against gram-positive and gram-negative microorganisms. Lasers in Surgery and Medicine. 2001;**29**:165-173. DOI: 10.1002/lsm.1105

[51] Hamblin MR, O'Donnell DA, Murthy N, Rajagopalan K, Michaud N, Sherwood ME, et al. Polycationic photosensitizer conjugates: effects of chain length and Gram classification on the photodynamic inactivation of bacteria. The Journal of Antimicrobial Chemotherapy. 2002;**49**:941-951. DOI: 10.1093/jac/dkf053

[52] Lehner R, Wang X, Marsch S, Hunziker P. Intelligent nanomaterials for medicine: carrier platforms and targeting strategies in the context of clinical application. Nanomedicine. 2013;**9**: 742-757. DOI: 10.1016/j.nano.2013.01.012

[53] Ni K, Luo T, Nash GT, Lin W. Nanoscale Metal-Organic Frameworks for Cancer Immunotherapy. Accounts of Chemical Research. 2020;**53**:1739-1748. DOI: 10.1021/acs.accounts.0c00313

[54] Li L, Huh KM. Polymeric nanocarrier systems for photodynamic therapy. Biomaterials Research. 2014;**18**: 19. DOI: 10.1186/2055-7124-18-19

[55] Gibot L, Lemelle A, Till U, Moukarzel B, Mingotaud AF, Pimienta V, et al. Polymeric micelles encapsulating photosensitizer: structure/photodynamic therapy efficiency relation. Biomacromolecules. 2014;**15**: 1443-1455. DOI: 10.1021/bm5000407

[56] Bechet D, Couleaud P, Frochot C, Viriot ML, Guillemin F, Barberi-Heyob M. Nanoparticles as vehicles for delivery of photodynamic therapy agents. Trends in Biotechnology. 2008;**26**:612-621. DOI: 10.1016/j.tibtech.2008.07.007

[57] Calixto GM, Bernegossi J, de Freitas LM, Fontana CR, Chorilli M. Nanotechnology-based drug delivery systems for photodynamic therapy of cancer: A review. Molecules. 2016;**21**: 342. DOI: 10.3390/molecules21030342

[58] Huang YY, Sharma SK, Dai T, Chung H, Yaroslavsky A, Garcia-Diaz M, et al. Can nanotechnology potentiate photodynamic therapy? Nanotechnology Reviews. 2012;**1**:111-146. DOI: 10.1515/ntrev-2011-0005

[59] Schwiertz J, Wiehe A, Gräfe S, Gitter B, Epple M. Calcium phosphate nanoparticles as efficient carriers for photodynamic therapy against cells and bacteria. Biomaterials. 2009;**30**:

3324-3331. DOI: 10.1016/j.biomaterials.2009.02.029

[60] Lucky SS, Soo KC, Zhang Y. Nanoparticles in photodynamic therapy. Chemical Reviews. 2015;**115**:1990-2042. DOI: 10.1021/cr5004198

[61] Deda DK, Araki K. Nanotechnology, light and chemical action: an effective combination to kill cancer cells. Journal of the Brazilian Chemical Society. 2015; **26**:2448-2470

[62] Foote CSM, of photosensitized oxidation. There are several different types of photosensitized oxidation which may be important in biological systems. Science. 1968;**162**(3857):963-970

[63] Fritsch C, Ruzicka T. Fluorescence diagnosis and photodynamic therapy in dermatology from experimental state to clinic standard methods. Journal of Environmental Pathology, Toxicology and Oncology: Official Organ of the International Society for Environmental Toxicology and Cancer. 2006;**25**(1–2): 425-439

[64] Moan JE. Properties for optimal PDT sensitizers. Journal of Photochemistry and Photobiology. B, Biology. 1990;**5**(3–4):521-524

[65] Ochsner MIF. Photophysical and photobiological processes in the photodynamic therapy of tumours. Journal of Photochemistry and Photobiology. B, Biology. 1997;**39**(1): 1-18

[66] Dolmans D, Fukumura D, Jain RK. Photodynamic therapy for cancer. Nature Reviews Cancer. 2003;**3**:380-387

[67] Juarranz Á, Jaen P, Sanz-Rodríguez F, Cuevas J, González S. Photodynamic therapy of cancer. Basic principles and applications. Clinical and Translational Oncology. 2008;**10**: 148-154

[68] Niculescu A-G, Grumezescu AM. Photodynamic Therapy—An Up-to-Date Review. Applied Sciences. 2021;**11**:3626

[69] Fisher AMR, Murphree AL, Gomer CJ. Clinical and preclinical photodynamic therapy. Lasers in Surgery and Medicine. 1995;**17**:2

[70] Braathen LR. Photodynamic therapy: When and why? Journal of the European Academy of Dermatology and Venereology. 2002;**16**:1063

[71] Fialová P, Vašků, V. PDT and its possible use in dermatology. Dermatol. praxi. 2017;**11**:166

[72] Ben-hur E, Horowitz B. Advances in photochemical approaches for blood sterilization. Photochemistry and Photobiology. 1995;**62**:383

[73] Wyss P, Tadir Y, Tromberg BJ, Haller U, editors. Photomedicine in Gynecology and Reproduction. S. Karger AG; 2000. DOI: 10.1159/isbn.978-3-318-00453-3. ISBN (print): 978-3-8055-6905-7. ISBN (electronic): 978-3-318-00453-3

[74] Van Vugt DA, Krzemien A, Roy BN, et al. Photodynamic Endometrial Ablation in the Nonhuman Primate. Reprod. Sci. 2000;**7**:125-130. DOI: 10.1177/107155760000700208

Chapter 3

Relevance of Oxidoreductases in Cellular Metabolism and Defence

Panchashree Das and Priyabrata Sen

Abstract

Oxidoreducatses occupy one-third of all enzymatic activities registered in the BRaunschweig ENzyme DAtabase (BRENDA). This group of enzymes are playing a vital role in plant growth and metabolism. Oxidoreducatses (EC 1) is the largest class of enzyme that includes dehydrogenases, oxygenase, peroxidise, oxidases and other enzymes that catalyse oxidation–reduction reaction by transferring electrons, hydrogen, or oxygen from a reductant molecule to an oxidant molecule. These enzymes play an important role in photosynthesis, aerobic and anaerobic respiration, amino acid metabolism and fatty acid metabolism. Besides metabolism these enzymes are also involve in providing defence against pathogens by activating signal transduction pathways. Here we have discussed in details about the sub-classes of oxidoreductase ezymes according to the reaction they catalyse and their importance in metabolism and defence against plant pathogen attack.

Keywords: oxidoreducatse, enzyme, metabolism, defence, pathogen attack

1. Introduction

Enzymes facilitate the biochemical reaction by lowering down the activation energy. In biochemistry, the enzymes are classified in six major classes according to the biochemical reaction it catalyses or the substrate it uses. Each class of the enzyme is named by the International Union of Biochemistry and Molecular Biology through EC (Enzyme Commission) number. Oxidoreductases are the enzymes that involved in different parts of natural lifecycle such as plants, animals and microorganisms. In plant biochemistry, oxidoreductase is the large group of enzymes having EC 1 nomenclature and it catalyses the oxidation and reduction reactions in plants. It mainly involves in transfer of electron molecule from one molecule (electron donor or reductant) to another (electron acceptor or oxidant). The oxidoreductases includes oxidase, oxygenase, peroxidase, dehydrogenase and other enzymes that catalyse biochemical reactions that involves in insertion of oxygen, transfer of hydride ions, transfer of electrons and protons [1]. These enzymes mainly utilise FAD (flavin adenine dinucleotide), Fd (ferredoxin) FMN (flavin mononucleotide), NAD (nicotinamide adenine dinucleotide), NADP (nicotinamide adenine dinucleotide phosphate), Coenzyme B, Coenzyme Q etc. for catalysing biochemical reactions in living cells [2]. Plants mainly use this class of enzymes for the metabolism process such as synthesis or degradation of biomolecules to sustain throughout the life span. In case of plants these enzymes also involves in metabolism of exogenous molecules such as herbicides [3]. The oxidoreductases have important roles in aerobic as well as anaerobic respiration, photosynthesis, electron transport chain, pentose phosphate pathway and photorespiration [4]. These enzymes also involve in amino acid metabolism. Oxidoreductases present in plant plasma membrane involves in maintaining the redox potential through proton pumping, nutrient uptake regulation, regulation of ion channels, reduction of iron and regulation of ion channels. Some oxidoreductase enzymes are also involved in signal transduction pathways and poses defence against pathogen attack.

This large group of enzyme comprises with number of enzymes that catalyses the transmission of electron from a donor (reduction) to an acceptor (oxidation) molecules with the help of cofactors to facilitate metabolism and provide defence against pathogens in plants (**Table 1**) [5]. In this review we mainly focused on the classification of oxidoreductase enzymes and their role in metabolism and defence in plants.

EC number	Description	Example
EC 1.1	Act on the CH-OH group of donors- The enzymes act on primary alcohols, secondary alcohols and hemi-acetals	Alcohol dehydrogenase, mannitol dehydrogenase, glucose 1-dehydrogenase, glycerol-3-phosphate dehydrogenase, phosphogluconate dehydrogenase
EC 1.2	Act on the aldehyde or oxo group of donors- The enzymes oxidise aldehydes to the corresponding acid. The oxo group of aliphatic compounds can be oxidised by addition of water and cleavage of a carbon–carbon bond and for aromatic compounds it can be done by addition of the elements of water and dehydrogenation.	Aldehyde dehydrogenase, benzaldehyde dehydrogenase, aspartate-semialdehyde dehydrogenase, pyruvate dehydrogenase, pyruvate synthase, aldehyde ferredoxin oxidoreductase
EC 1.3	Act on the CH-CH group of donors- The enzyme catalyses chemical reactions that introduce a double-bond into the substrate by direct dehydrogenation at a carbon–carbon single bond	Dihydropyrimidine dehydrogenase, L-galactonolactone dehydrogenase, fumarate reductase, coproporphyrinogen oxidase
EC 1.4	Act on the CH-NH_2 group of donors- It includes mainly the amino-acid dehydrogenases and the amine oxidases	Alanine dehydrogenase, D-arginine dehydrogenase, valine dehydrogenase
EC 1.5	Act on the CH-NH group of donors- It contains enzymes that dehydrogenate secondary amines	Pyrroline-5-carboxylate reductase, D-lysopine dehydrogenase, trimethylamine dehydrogenase
EC 1.6	Act on NADH or NADPH- This subclass of enzymes utilise NADH or NADPH to reduce a substrate.	NAD(P) + transhydrogenase, leghemoglobin reductase, NAD(P)H oxidase, demethylphylloquinone reductase
EC 1.7	Act on other nitrogenous compounds as donors- The enzymes are involved in oxidising various nitrogenous substrates.	Nitrate reductase, azobenzene reductase, nitroalkane oxidase, acetylindoxyl oxidase, hydroxylamine reductase
EC 1.8	Act on a sulphur group of donors- The enzymes react on inorganic substrates or organic thiols.	Assimilatory sulphite reductase, cystine reductase, thioredoxin-disulfide reductase, adenylyl-sulfate reductase
EC 1.9	Act on a heme group of donors- It includes enzymes that act on heme group of cytochrome.	Cytochrome-c oxidase, nitrate reductase

EC number	Description	Example
EC 1.10	Act on diphenols and related substances as donors- It includes the enzyme4s that involve in oxidation of diphenols or ascorbate	trans-Acenaphthene-1,2-diol dehydrogenase, catechol oxidase, L-ascorbate oxidase, ubiquinol oxidase
EC 1.11	Act on a peroxide as acceptor- It includes peroxygenases and peroxidases	NADH peroxidise, catalase, glutathione peroxidise, fatty-acid peroxygenase
EC 1.12	Act on hydrogen as donor- It includes hydrogenases aside from those that use iron–sulphur compounds as donor to reduce H^+ to H_2	Hydrogen dehydrogenase, cytochrome-c3 hydrogenase, hydrogen:quinone oxidoreductase, hydrogenase
EC 1.13	Act on single donors with incorporation of molecular oxygen- It contains oxygenases that catalyses the incorporation of molecular oxygen to the substrate	Catechol 1,2-dioxygenase, gentisate 1,2-dioxygenase, tryptophan 2,3-dioxygenase, indole 2,3-dioxygenase, tryptophan 2′-dioxygenase
EC 1.14	Act on paired donors, with incorporation or reduction of molecular oxygen- The subclass includes enzymes that act on two hydrogen donors and oxygen is incorporate into one or both of the donors during biochemical reaction.	Pyrimidine-deoxynucleoside 2′-dioxygenase, thymine dioxygenase, gibberellin 3 $_{\beta}$-dioxygenase, L-isoleucine 4-hydroxylase
EC 1.15	It contains enzymes acting on superoxide as acceptor with a single sub-subclass	Superoxide dismutase, superoxide reductase
EC 1.16	Oxidising metal ions- The enzymes of this subclass involve in oxidising metal ions to higher valency state	Mercury(II) reductase, aquacobalamin reductase
EC 1.17	Act on CH or CH_2 groups- The enzymes of this subclass involve in oxidative conversion of the -CH_2- group of donors to -CHOH- (or -CH- to -COH-) and sugars to deoxysugars.	Leucoanthocyanidin reductase, xanthine dehydrogenase, nicotinate dehydrogenase, xanthine oxidase
EC 1.18	It contains enzymes acting on iron–sulphur proteins as donors and having receptors like NAD+ or NADP+ and dinitrogen.	Ferredoxin—NADP+ reductase, vanadium-dependent nitrogenise
EC 1.19	It contains enzymes on reduced flavodoxin as donors and having receptors like NAD+ or NADP+ and dinitrogen	Flavodoxin—NADP+ reductase, nitrogenase (flavodoxin)
EC 1.20	Act on phosphorus or arsenic in donors	Phosphonate dehydrogenase, arsenate reductase, methylarsonate reductase
EC 1.21	Catalysing the reaction X-H + Y-H = X-Y- Catalyse the reaction by forming or breaking an X-Y bond	Iodotyrosine deiodinase, isopenicillin-N synthase, D-proline reductase
EC 1.22	Act on halogen donors	Iodotyrosine deiodinase
EC 1.23	Catalyses the reaction by reducing C-O-C group as acceptor	(+)-Pinoresinol reductase, violaxanthin de-epoxidase
EC 1.97	This subclass includes other oxidoreductases that are not included in previous groups	Chlorate reductase, pyrogallol hydroxytransferase

Table 1.
Classification of oxidoreducatse enzymes.

2. Classification of oxidoreductase

Oxidoreductase is having biological importance due to its abundance in the living organisms. This large class of enzyme is classified in several subclasses according to the reaction it catalyses, the three dimensional structure. These enzymes can be either oxidases or dehydrogenases. The oxidases utilises molecular oxygen as electron or hydrogen acceptor and dehydrogenases acts as electron or hydrogen donor for the NAD+/NADP+ or a flavin enzymes. The other subclasses of oxidoreductases include peroxidases, hydroxylases, oxygenases, and reductases. Peroxidases catalyse the reduction of hydrogen peroxide in peroxisome. Hydroxylases mainly acts by adding hydroxyl groups to its substrates. Oxygenases involves in incorporating molecular oxygen into organic substrates. Reductases mostly act as oxidases involves in reduction reaction. All of the enzymes belonging to this class are playing essential role in plant lifecycle and stress signalling.

The oxidoreducatase enzymes are securing EC 1 in the enzyme classification given by the International Union of Biochemistry and Molecular Biology and it is having 24 sub-classes [6].

3. Oxidoreductase enzymes involved in metabolism

3.1 Oxidoreductase enzymes involved in photosynthesis

Photosynthesis is the most important biochemical process completed in chloroplast of plant cell for conversion of light energy into chemical energy. In this process the green plants absorbs light energy, water (H_2O) and carbon dioxide (CO^2) to form oxygen and energy-rich organic compound. The photosystem I (PS I) and photosystem II (PS II) are involved in the light reaction of photosynthesis. PS II liberates the hydrogen molecules from water at the time of photolysis and frees molecular oxygen. After that the liberated electrons in this process are transferred to plastoquinol, Cytochrome b_6 complex and finally to plastocyanin. At this point the PS I take electrons from plastocyanin and transfer it to iron-containing compound ferredoxin. Upon taking the electrons, ferredoin becomes reduced ferredoxin. The Ferredoxin-$NADP^+$ reductase, an enzyme belongs to the class oxidoreductase, catalyses the biochemical reaction of formation of NADPH from $NADP^+$(Nicotinamide adenine dinucleotide phosphate) during photosynthesis [7, 8]. The enzyme Ferredoxin-$NADP^+$ reductase transfer an electron from each of two reduced ferredoxin molecules to a single molecule of the two electron carrier NADPH and after completion of the reaction ferredoxin becomes oxidised (**Figure 1**). This FNR enzyme utilises a flavin cofactor FAD (flavin adenine dinucleotide), the iron–sulphur protein ferredoxin as donor and $NADP^+$ as acceptor.

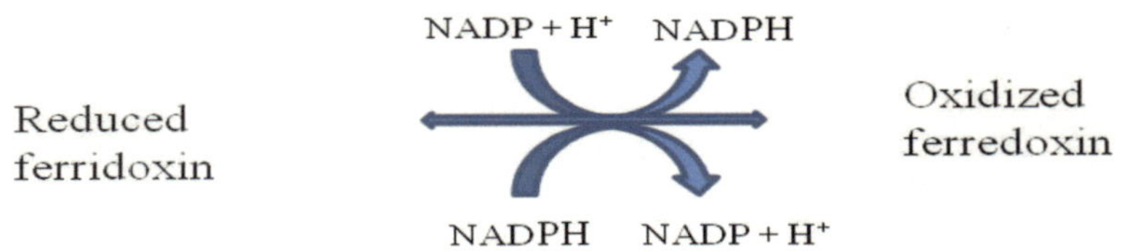

Figure 1.
Chemical reaction catalysed by FNR (Ferredoxin-$NADP^+$ reductase).

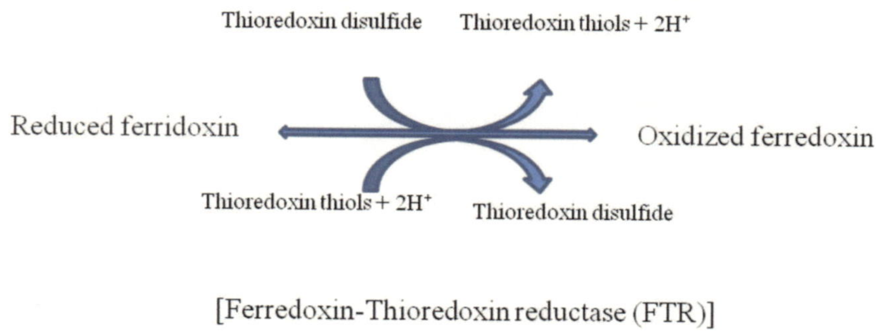

Figure 2.
Chemical reaction catalysed by FTR (Ferredoxin-Thioredoxin reductase).

Plants utilise Ferredoxin-Thioredoxin reductase (FTR), a member of oxidoreductase class of enzymes, to regulate the carbon fixation in photosynthetic organisms [9]. FTR helps the plants to regulate the carbohydrate metabolism based on availability of light. During day light the FTR reduces oxidised thioredoxin by reduced ferredoxin. This reduced thioredoxin activates certain carbohydrate synthesis enzymes such as chloroplast fructose-1,6-bisphosphatase, Sedoheptulose-bisphosphatase and phosphoribulokinase by cleaving disulfide bonds in enzymes [10]. This leads to carbohydrate synthesis during day time. At night, the ferredoxin remains oxidised and thus it cannot activate thioredoxin to promote carbohydrate biosynthesis and breakdown of carbohydrate starts (**Figure 2**).

3.2 Oxidoreductase enzymes involved in chlorophyll biosynthesis

The chlorophyll biosynthesis is a fundamental metabolic process as it is the most important pigment for photosynthesis. The enzyme protochlorophyllide oxidoreductase (POR), a member of oxidoreductase classes of enzyme, is a key regulatory enzyme for chlorophyll biosynthesis that catalyses the light-induced reduction of protochlorophyllide (PChlide) into chlorophyllide (Chlide) in the presence of NADPH (reduced form of NADP) [11–14]. The enzyme POR utilises NADPH (reduced nicotinamide adenine dinucleotide phosphate) as cofactor [11–14]. There are two or more isoforoms of this POR enzyme has been identified in different plant species. In barley (*Hordeum vulgare*), POR A and POR B, the two distinct isozymes of POR are present [15]. Arabidopsis (*Arabidopsis thaliana*) is having three distinct isozymes of POR such as POR A, POR B and POR C [16, 17].

3.3 Oxidoreductase enzymes involved in glycolysis

Glycolysis is the carbohydrate catabolism pathway that converts glucose into pyruvate and produces NADH (reduced nicotinamide adenine dinucleotide) and ATP (adenosine triphosphate). This process contains two phases, preparatory phase and pay-off phase. In the first phase, glucose breaks down to form two trios sugar. So in the second phase of glycolysis each reaction occurs twice for each glucose molecule. In the pay-off phase Glyceraldehyde 3-phosphate dehydrogenase catalyses the oxidative conversion of glyceraldehyde-3-phosphate to 1,3-bisphosphoglycerate (1,3-BPG) and reduces NAD^+ to NADH (**Figure 3a**). For maintaining the redox state of the cell there has to be a proper balance between NAD^+/NADH. For aerobic respiration the

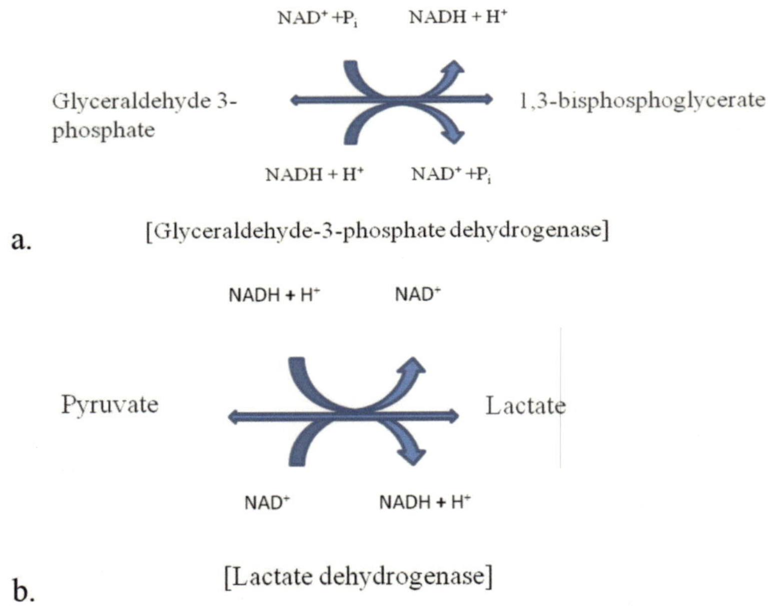

Figure 3.
Chemical reaction catalysed by oxidoreductase enzymes in glycolysis. (a) Chemical reaction catalysed by Glyceraldehyde 3-phosphate dehydrogenase. (b) Chemical reaction catalysed by lactate dehydrogenase.

oxidation of NADH occurs during tricarboxylic acid cycle (TCA) and for fermentation it occurs at the time of conversion of pyruvate to lactate by lactate dehydogenase enzyme (**Figure 3b**).

3.4 Oxidoreductase enzymes involved in TCA cycle (Tricarboxylic acid cycle)

Tricarboxylic acid cycle (TCA) consists of a series of enzyme catalysed biochemical reactions. The precursor of TCA cycle comes from carbohydrate, protein and fats [18]. The product of glycolysis, pyruvate converts in acetyl-CoA to enter in TCA cycle. Except leucin and lysine, the other amino acids degraded to TCA cycle intermediates. Trough beta-oxidation of fatty acid, it converted into acetyl-CoA that enters into TCA cycle. During TCA cycle, a large number of NADH and FADH molecules are produced by different oxidoreductase enzymes [19]. Pyruvate dehydrogenase catalyses the conversion of pyruvate to acetyl-CoA that enter in the TCA cycle (**Figure 4a**). Isocitrate dehydrogenase involves in oxidative decarboxylation of isocitrate to α-ketoglutarate (**Figure 4b**). Another TCA cycle enzyme α-ketoglutarate dehydrogenase catalyse the generation of succinyl-CoA from α-ketoglutarate through oxidative decarboxylation (**Figure 4c**). Succinate dehydrogenase involves in oxidative conversion of succinate to fumarate (**Figure 4d**). Malate dehydrogenase enzyme of TCA cycle catalyses the conversion of malate to oxaloacetate through oxidation (**Figure 4e**).

3.5 Oxidoreductase enzymes involved in ETC (electron transport chain) and Oxidative phosphorylation

Oxydative phosphorylation involves flow of electrons from a series of proteins and electron carriers within the *mitochondrial membrane by ETC.* Several biochemical reactions like glycolysis, TCA cycle, fatty acid oxidation and amino acid oxideation

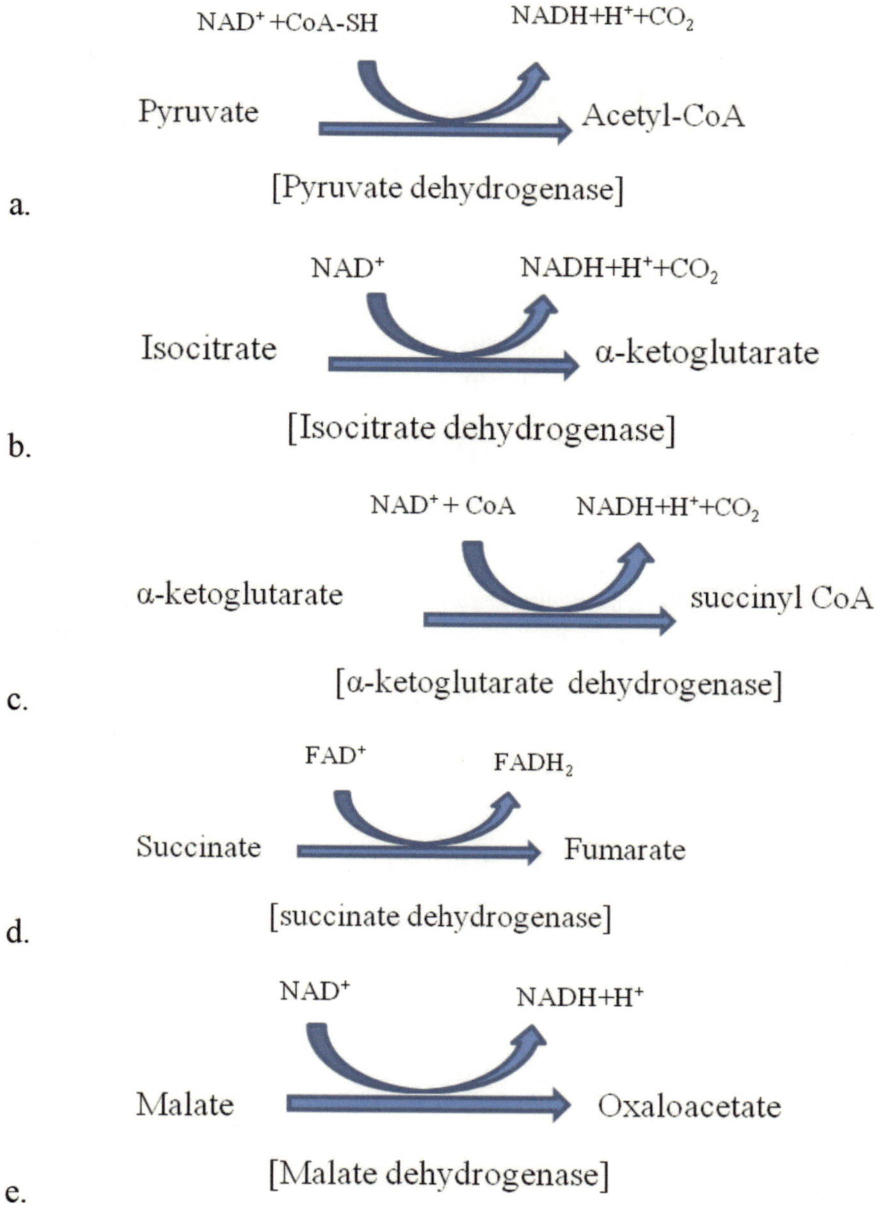

Figure 4.
Chemical reaction catalysed by oxidoreductase enzymes in TCA cycle. (a) Chemical reaction catalysed by pyruvate dehydrogenase. (b) Chemical reaction catalysed by isocitrate dehydrogenase. (c) Chemical reaction catalysed by α-ketoglutarate dehydrogen. (d) Chemical reaction catalysed by succinate dehydrogen. (e) Chemical reaction catalysed by malate dehydrogen.

yields high energy-rich molecules NADH (reduced nicotinamide adenine dinucleotide) and $FADH_2$ (reduced flavin adenine dinucleotide). This NADH and $FADH_2$ have pair of electron with high transfer potential. In ETC the electrons of NADH and $FADH_2$ are transferred to molecular oxygen through a series of electron carriers and the energy produced during this process leads to formation of ATP [20]. Different subclasses of oxidoreductase enzymes catalyse the oxidative phosphorylation

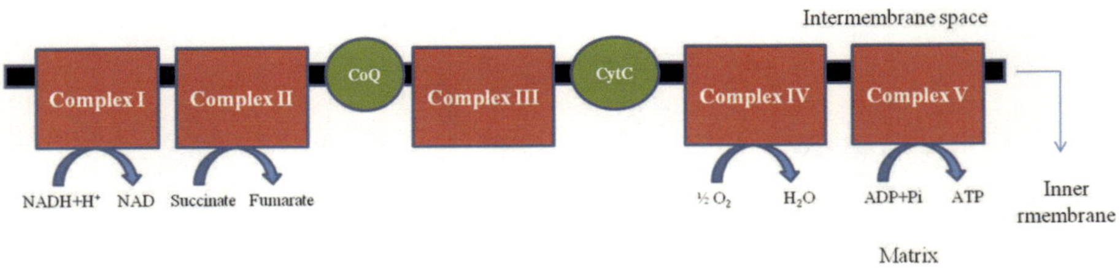

Figure 5.
Enzyme complex involved in ETC in mitochondria.

reactions (**Figure 5**). The electron transport chain is consists of four complexes. NADH:quinone oxidoreductase or NADH dehydrogenase (complex I) contains FMN that accepts electrons and H^+ from NADH and forms $FMNH_2$. It also contains iron–sulphur protein that helps in transfer of electrons and H^+ to Coenzyme Q. Succinate:quinine oxidoreductase or succinate dehydrogenase (complex II) contains iron and a bound FAD cofactor to oxidise succinate to fumarate and reduces ubiquinone. Coenzyme Q collects the electrons from complex I and Complex II and transfers them to complex III. Quilon:cytochrome c oxidoreductase (complex III) contains heme group, in which Fe^{3+} accepts electrons from Coenzyme Q (CoQ) to form Fe^{2+} and transfers the electrons to cytochrome C (CytC), that transfers the electrons to complex IV. Cytochrome C:oxygen oxidoreductase, an aa3-type enzyme (complex IV) contains heme group in which Fe^{3+} accepts electrons from cytochrome C to form Fe^{2+} and transfers electrons to oxygen that combines with hydrogen to form water molecule [21].

3.6 Oxidoreductase enzymes involved in amino acid metabolism

Amino acid dehydrogenases are involved in amino acid catabolism reaction [22]. This enzyme utilises $NAD(P)^+$ as cofactor. Amino acid dehydrogenases catalyse the biochemical reaction by transferring the hydride from the Cα atom of an amino acid to $NAD(P)^+$ and forms intermediate α-imino acid that finally hydrolyzed to α-keto acid and ammonium (**Figure 6**). Glutamate dehydrogenase

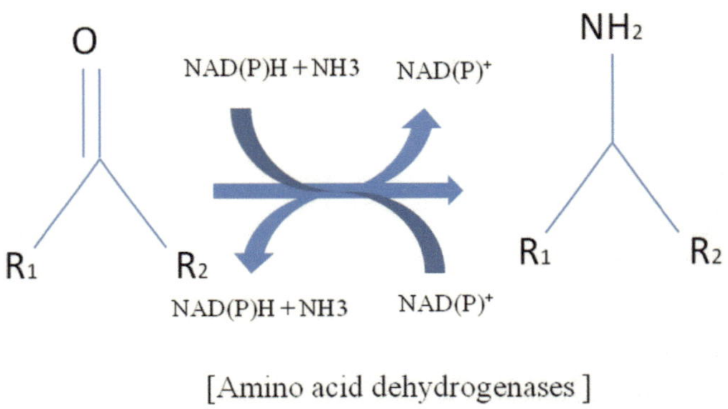

Figure 6.
Chemical reaction catalysed by amino acid dehydrogenase.

and phenylalanine dehydrogenase are involved in oxidative deamination reaction for amino acid catabolism [23, 24]. Glutamate dehydrogenase and phenylalanine dehydrogenase catalyse deamination reaction of glutamate and phenylalanine respectively [23, 24].

3.7 Oxidoreductase enzymes involved in fatty acid metabolism

Oxidoreductase enzymes are involved in metabolism of fatty acids, especially in β-oxidation. The β-oxidation procedure of fatty acid catabolism occurs in cytoplasm of cell. Activation of acyl-CoA by acyl-CoA synthetases is seen to initiate β-oxidation in the membrane of the peroxisome [25–27]. The β-oxidation procedure of fatty acid catabolism occurs in cytoplasm of cell. The products of this pathway are acetyl-CoA, NADH and FADH. The pathway needs enzymes like acyl-CoA dehydrogenase and 3-hydroxyacyl-CoA dehydrogenase that belongs to the oxidoreductase classes of enzyme [28]. Acyl-CoA dehydrogenases (ACAD) catalyse the oxidation of acyl-CoA to trans-2-enoyl-CoA with the help of FAD [25]. β-ketoacyl-ACP reductase and enoyl-ACP reductase are involved in *de novo* synthesis of fatty acid synthase are belongs to oxidoreductase class of enzymes [29]. NAD^+ oxidoreductase is involved in α-oxidation in plants [30]. Acyl-CoA oxidases (ACOX) catalyse the oxidation of acyl-CoA to trans-2-enoyl-CoA [25].

4. Oxidoreductase enzymes involved in plant defence mechanism

The oxidoreductases involved in defence mechanism of plants are typically helping in ROS (reactive oxygen species) generation. In reaction to stressors, ROS serve as signalling molecules and activate signal transduction pathways. Accumulation of ROS (reactive oxygen species) at the site of pathogen infection leads to hypersensitive response that causes death of plant tissue at the site of infection to restrict further spread of pathogen infection [31, 32]. The pathogenesis-related (PR) genes are expressed more frequently as part of the hypersensitive response (HR), which also comprises the production of antimicrobial secondary metabolites and a type of localised cell death (LCD) at the infection site. Since electron utilisation in the chloroplast stroma shuts down during the HR, the photosynthetic electron transport chain undergoes excessive reduction, which is the initial contributor to ROS. The generation of ROS molecules are catalysed by NADPH-dependent oxidase system. In plants this system generates superoxide anion (O_2^-) and hydrogen peroxide (H_2O_2) in response to pathogen attack [33, 34]. Like NADPH-dependent oxidases, Germin-like oxalate oxidases, glycolate oxidases, Amine oxidases, class III peroxidise which belong to oxidoreductase class of enzymes are involve in ROS generation in response to pathogen attack [35–39]. All the enzymes associated with ROS production belongs to the class of oxidoreductase of enzyme [40].

The ROS are generated by cell organelles like chloroplast, mitochondria and peroxisome as by-product of aerobic respiration. The main parts of cells that are involved into generation of H_2O_2 are Cytoplasm, plasma membrane, apoplasts, endoplasmic reticulum, and extracellular matrix. H_2O_2 is a by-product of several metabolic pathways like glycolate oxidase reaction, fatty acid β-oxidation, electron transport chain, oxidative phosphorylation, transition metal ions, thymidine, and polyamine catabolism etc. [41–44]. Enzymatic activities of plasma-membrane-localised NADPH oxidases, amine oxidases, and cell wall peroxidises are playing major role in ROS production [45].

Reactive oxygen species (ROS) and reactive nitrogen species (RNS) are both significant signals in plants and vital regulators of a number of activities, including metabolism, growth and development, response to abiotic and biotic stressors, solute transport, autophagy, and programmed cell death (PCD). ROS plays a crucial role as signalling molecule in defence mechanism as well as in developmental process in all plants [46–48]. During biotic stress and abiotic stress condition, the cellular homeostasis of plants is disrupted and this lead towards higher production ROS [45, 46]. Increasing concentration of ROS can become a threat for living cell as it can lead to cell death. ROS may damage nucleic acids, proteins and activate apoptasis. So, high level of co-ordination between signalling and metabolic pathways for producing and scavenging ROS is needed during a stress condition is needed [44, 45].

The plant-pathogen interaction also involves generation of reactive nitrogen species (RNS) (NO, NO^+, NO^-, NO_2 and $ONOO^-$) at the site of infection [35]. This also acts as a signalling molecule during stress condition. ROS and RNS together involve in hypersensitive response as well as programmed cell death responses during pathogen infection [49].

Under normal condition, cellular concentration of ROS and RNS is low in plants. But during stress condition, both ROS and RNS act as a signalling molecule to activate plant defence responses against the stress. The rapid production of ROS develops oxidative stress to the plants. A high concentration of ROS and RNS can lead to cell death. To minimise the toxic action of ROS, plants posses antioxidant activities [33]. It is very much needed to maintain the equilibrium between production and metabolism of ROS and RNS. The detoxification of these substances is done enzymatic as well as non-enzymatic antioxidants [42]. Each and every cell compartments are having their own mechanism of scavenging excess amount of ROS and RNS.

The nonenzymatic antioxidants like ascorbate, glutathione, tocopherol etc. can directly detoxify ROS or reduce substrate for antioxidant enzymes. Ascorbate is an important compound to directly eliminate singlet oxygen and super oxide ions. Ascorbate also involves in reduction of H_2O_2 to water through ascorbate–glutathione cycle [50]. ROS oxidised glutathione to form oxidised glutathione. Glutathione and oxidised glutathione helps in maintaining redox balance of cellular components [42, 51]. α-tocopherol detoxifies peroxyl ions (HO_2) in lipid bilayer and chloroplast membrane [52].

The enzymatic antioxidants like superoxide dismutase, catalase, peroxiredoxins and peroxidases are involved in ROS degradation. Superoxide dismutase catalyses the formation of molecular oxygen and hydrogen peroxide from super oxide anions. Catalase detoxifies H_2O_2 to water. The enzymes involved in ascorbate–glutathione cycle, i.e. glutathione reductase (GR), dehydroascorbate reductase (DHAR), monohydroascorbate reductase (MDAR) and ascorbate peroxidise (APX) help in generating soluble antioxidants [53]. In plants, the most important ROS scavenging mechanism is ascorbate–glutathione cycle [42].

The oxidoreductase enzymes involved in plant growth and defence against stress are enlisted in **Tables 2** and **3**.

Sl. no.	Plants	Gene	Stress condition	Reference
1.	Tea (*Camellia sinensis*)	*glutathione S-transferase*	pathogen and insect attack, cold spells, drought and salt stresses, nitrogen nutrition	[54]
2.	*Triticum aestivum*	*glutathione reductase* (*GR*)	drought, heat, salt and arsenic stress	[55]
3.	Arabidopsis	CRL1 (Cinnamoyl coA: NADP oxidoreductase-like 1)	drought stress	[56]
4.	pepper (*Capsicum annuum*)	CaOXR1 (*C. annuum* oxidoreductase protein)	Salt and oxidative stress	[57]
5.	pea (*Pisum sativum* L.)	xanthine oxidase (XOD)	abiotic stress by heavy metals	[58]
6.	Arabidopsis	Glutaredoxins (GRXs)	stress responses	[59]
7.	Arabidopsis	electron transfer flavoprotein/ electron-transfer flavoprotein: ubiquinone oxidoreductase (ETF/ETFQO)	carbohydrate starvation	[60]

Table 2.
Involvement of oxidoreductases under stress condition in plants.

Sl no.	Enzyme name	Role	Reference
1.	flavoprotein:ubiquinone oxidoreductase (ETF/ETFQO) complex	Seed development and germination (Arabidopsis)	[61]
2.	CRL1 (Cinnamoyl coA: NADP oxidoreductase-like 1)	seed germination and flowering	[56]
3.	Glutaredoxins (GRXs)	floral development, organ identity gene expression, regulation of organ primordia initiation meiosis progression in the male germ line	[59]
4.	thiol-disulfide oxidoreductase PDI1	regulates actin structures in *Oryza sativa* root cells	[62]

Table 3.
Involvement of oxidoreductases in plant growth and development.

5. Conclusion

Oxidoreductase is the largest class of enzyme that includes several enzymes having importance in plant metabolism as well as defence mechanism. Number of metabolic pathways like chlorophyll biosynthesis, photosynthesis, glycolysis, fatty acid metabolism, tricarboxylic acid cycle includes oxidoreductase enzymes that utilises NAD, FAD, or NADP as a cofactor (**Figure 7**). This large class of enzyme contributes in developing plant defence responses against biotic and abiotic stress. Antioxidant enzymes maintain redox balance in cellular components in stress condition.

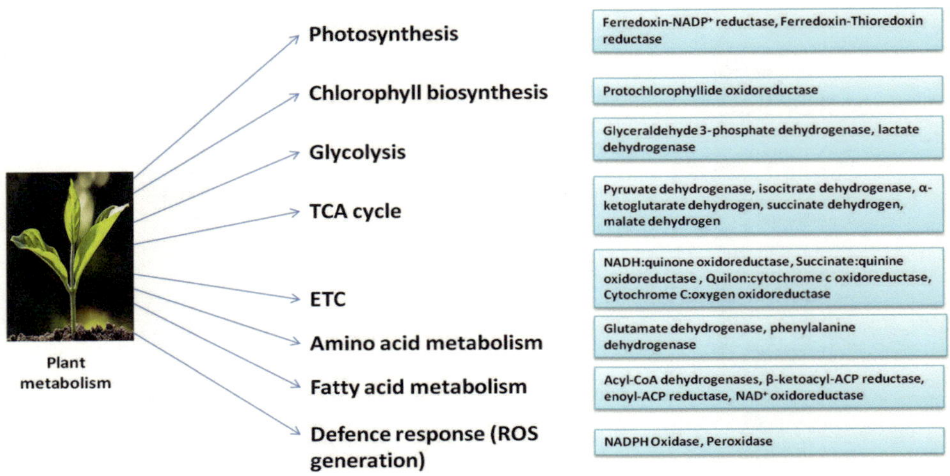

Figure 7.
Summary of role of different oxidoresductases in plant metabolism and defence.

Acknowledgements

PD would like to acknowledge Department of Agricultural Biotechnology, Assam Agricultural University and Centurion University of Technology and Management for the support. PS would like to acknowledge Department of Agricultural Biotechnology, Assam Agricultural University for the support.

Conflict of interest

All authors declare that they have no conflict of interest.

Data availability and material

Raw data are available upon request.

Ethics approval

As no human or mammalian subjects were involved in this research, no ethics approvals were required for this study.

Informed consent

All authors consent to participate in publication of these data.

Consent for publication

All authors consent to publish this article.

Author details

Panchashree Das[1]* and Priyabrata Sen[2]

1 Department of Biotechnology, MSSSoA, Centurion University of Technology and Management, India

2 Department of Agricultural Biotechnology, Assam Agricultural University, Jorhat, India

*Address all correspondence to: panchashreedas@gmail.com

References

[1] Trisolini L, Gambacorta N, Gorgoglion R. FAD/NADH dependent oxidoreductases: From different amino acid sequences to similar protein shapes for playing an ancient function. Journal of Clinical Medicine. 2019;**8**:2117

[2] McDonald A. The enzyme list class 1-Oxidoreductases. Nomenclature Committee of the International Union of biochemistry and molecular biology (NC-IUBMB). 2019

[3] Cummins I, Dixon DP, Freitag-Pohl S, Skipsey M, Edwards R. Multiple roles for plant glutathione transferases in xenobiotic detoxification. Drug Metabolism Reviews. 2011;**43**:266-280

[4] Geigenberger P, Fernie AR. Metabolic control of redox and redox control of metabolism in plants. Antioxidants & Redox Signaling. 2014;**21**:1389-1421

[5] Nicholas C, Lewis S. Fundamentals of Enzymology: The Cell and Molecular Biology of Catalytic Proteins. Oxford: Oxford University Press; 1999

[6] Price NC, Stevens L. The purification of enzymes. In: Fundamentals of Enzymology: The Cell and Molecular Biology of Catalytic Proteins. USA: Oxford University Press; 1999. pp. 15-46

[7] Berg JM, Tymoczko JL, Stryer L. Biochemistry. 6th ed. New York: W.H. Freeman; 2007

[8] Berg JM, Tymoczko JL, Stryer L. Biochemistry. 5th ed. New York: Freeman; 2002

[9] Masakazu H, Peter S, Jean-Pierre J. Oxidation-reduction properties of chloroplast tioredoxins, ferredoxin: Thioredoxin reductase, and thioredoxin f-regulated enzymes. Biochemistry. 1999;**38**:5200-5205

[10] Jacquot J, Lancelin J, Meyer Y. Thioredoxins: Structure and function in plant cells. New Phytologist. 1997;**136**:543-570

[11] Garrone A, Archipowa N, Zipfel PF, Hermann G, Dietzek B. Plant Protochlorophyllide oxidoreductases A and B: Catalytic efficiency and initial reaction steps. The Journal of Biological Chemistry. 2015;**290**:28530-28539

[12] Heyes DJ, Hunter CN. Making light work of enzyme catalysis: Protochlorophyllide oxidoreductase. Trends in Biochemical Sciences. 2005;**30**:642-649

[13] Reinbothe C, El Bakkouri M, Buhr F, Muraki N, Nomata J, Kurisu G, et al. Chlorophyll biosynthesis: Spotlight on protochlorophyllide reduction. Trends in Plant Science. 2010;**15**:614-624

[14] Townley HE, Sessions RB, Clarke AR, Dafforn TR, Griffiths WT. Protochlorophyllide oxidoreductase: A homology model examined by site-directed mutagenesis. Proteins. 2001;**44**:329-335

[15] Reinbothe S, Reinbothe C, Lebedev N, Apel K. PORA and PORB, two light-dependent protochlorophyllide reducing enzymes of angiosperm chlorophyll biosynthesis. Plant Cell. 1996;**8**:763-769

[16] Pattanayak GK, Tripathy BC. Catalytic function of a novel protochlorophyllide oxidoreductase C of *Arabidopsis thaliana*. Biochemical and Biophysical Research Communications. 2002;**291**:921-924

[17] Su Q, Frick G, Armstrong G, Apel K. PORC of *Arabidopsis thaliana*: A third light- and NADPH-dependent protochlorophyllide oxidoreductase that is differentially regulated by light. Plant Molecular Biology. 2001;**47**:805-813

[18] Mailloux RJ. Still at the center of it all; novel functions of the oxidative Krebs cycle. Bioenergetics. 2015;**4**:1-14

[19] Ponizovskiy MR. Role of Krebs cycle in the mechanism of stability internal medium and internal energy in an organism in norm and in mechanisms of cancer pathology. Modified Chemical Applications. 2016;**4**:191

[20] Mitchell P, Moyle J. Chemiosmotic hypothesis of oxidative phosphorylation. Nature. 1967;**213**:137-139

[21] Matsubayashi M, Inaoka DK, Komatsuya K. Novel characteristics of mitochondrial electron transport chain from Eimeria tenella. Genes. 2019;**10**:29

[22] Vidal LS, Kelly CL, Mordaka PM, Heap JT. Review of NAD(P)H-dependent oxidoreductases: Properties, engineering and application. BBA – Proteins and Proteomics. 2018;**1866**:327-347

[23] Asano Y, Nakazawa A, Endo K. Novel phenylalanine dehydrogenases from *Sporosarcina ureae* and *Bacillus sphaericus* purification and characterization. The Journal of Biological Chemistry. 1987;**262**:10346-10354

[24] Frieden C. Glutamate dehydrogenase. VI. Survey of purine nucleotide and other effects on the enzyme from various sources. International Journal of Biological Chemistry. 1965;**240**:2028-2035

[25] Quijano C, Trujillo M, Castro L, Trostchansky A. Interplay between oxidant species and energy metabolism. Redox Biology. 2016;**8**:28-42. DOI: 10.1016/j.redox.2015.11.010

[26] Reddy JK, Hashimoto T. Peroxisomal beta-oxidation and peroxisome proliferator-activated receptor alpha: An adaptive metabolic system. Annual Review of Nutrition. 2001;**21**:193-230

[27] Suzuki H, Kawarabayasi Y, Kondo J, Abe T, Nishikawa K, Kimura S. Structure and regulation of rat long-chain acyl-CoA synthetase. The Journal of Biological Chemistry. 1990;**265**(15):8681-8685

[28] Voet D, Voet JG, Pratt CW. Fundamentals of Biochemistry: Life at the Molecular Level. 4th ed. Hoboken, NJ: Wiley; 2013

[29] Harwood JL. Plant Fatty Acid Synthesis. Urbana, IL: The AOCS Lipid Library; 2010

[30] Saffert A, Hartmann-Schreier J, Schön A, Schreier P. A dual function α-dioxygenase-peroxidase and NAD+ oxidoreductase active enzyme from germinating pea rationalizing α-oxidation of fatty acids in plants. Plant Physiology. 2000;**123**(4):1545-1552

[31] Heath MC. Hypersensitive response-related death. Plant Molecular Biology. 2000;**44**:321-334

[32] Lamb C, Dixon RA. The oxidative burst in plant disease resistance. Annual Review of Plant Physiology and Plant Molecular Biology. 1997;**48**:251-275

[33] Apel K, Hirt H. Reactive oxygen species: Metabolism, oxidative stress, and signal transduction. Annual Review of Plant Biology. 2004;**55**:373-399

[34] Asada K, Takahashi M. Production and scavenging of active oxygen in photosynthesis. In: Kyle DJ, Osmond CB,

Arntzen CJ, editors. Photoinhibition. Amsterdam: Elsevier; 1987. pp. 227-287

[35] Almagro L, Gómez Ros LV, Belchi-Navarro S, Bru R, Ros Barceló A, Pedreño MA. Class III peroxidases in plant defence reactions. Journal of Experimental Botany. 2009;**60**:377-390

[36] Hu X, Bidney DL, Yalpani N, Duvick JP, Crasta O, Folkerts O, et al. Overexpression of a gene encoding hydrogen peroxide-generating oxalate oxidase evokes defense responses in sunflower. Plant Physiology. 2003;**133**:170-181

[37] Mika A, Minibayeva F, Beckett R, Luthje S. Possible functions of extracellular peroxidases in stress-induced generation and detoxification of active oxygen species. Phytochemistry Reviews. 2004;**3**:173-193

[38] Mittler R, Vanderauwera S, Gollery M, Breusegem FV. Reactive oxygen gene network of plants. Trends in Plant Science. 2004;**9**:490-498

[39] Walters DR. Polyamines and plant disease. Phytochemistry. 2003;**64**:97-107

[40] André-Lévigne D, Modarressi A, Pepper MS, Pittet-Cuénod B. Reactive oxygen species and NOX enzymes are emerging as key players in cutaneous wound repair. International Journal of Molecular Sciences. 2017;**18**(10):2149. DOI: 10.3390/ijms18102149

[41] Amendola R, Cervelli M, Tempera G. Spermine metabolism and radiation-derived reactive oxygen species for future therapeutic implications in cancer: An additive or adaptive response. Amino Acids. 2014;**4**:487-498

[42] Caverzan A, Casassola A, Brammer SP. Reactive oxygen species and antioxidant enzymes involved in plant tolerance to stress. In: Shanker AK, Shanker C, editors. Abiotic and Biotic in Stress Plants - Recent Advances Future Perspectives. London, UK, London: InTechOpen; 2016. pp. 463-480

[43] Kudryavtseva AV, Krasnov GS, Dmitriev AA. Mitochondrial dysfunction and oxidative stress in aging and cancer. Oncotarget. 2016;7:44879-44905

[44] Sharma P, Jha AB, Dubey RS, Pessarakli M. Reactive oxygen species, oxidative damage, and antioxidative defense mechanism in plants under stressful conditions. Journal of Botany. 2012;**2012**:1-26

[45] Racchi ML. Antioxidant defenses in plants with attention to Prunus and Citrus spp. Antioxidants. 2013;**2**:340-369

[46] Dowling DK, Simmons LW. Life-history evolution reactive oxygen species as universal constraints in life history. Proceedings of the Biological Sciences. 2009;**276**:1737-1745

[47] Grassmann J, Hippeli S, Elstner EF. Plant's defence and its benefits for animals and medicine: Role of phenolics and terpenoids in avoiding oxygen stress. Plant Physiology and Biochemistry. 2002;**40**:471-478

[48] Hippeli S, Elstner EF. Mechanisms of oxygen activation during plant stress: Biochemical effects of air pollutants. Journal of Plant Physiology. 1996;**148**:249-257

[49] Neill SJ, Desikan R, Hancock JT. Nitric oxide signaling in plants. New Phytologist. 2003;**159**:11-35

[50] Foyer CH, Noctor G. Ascorbate and glutathione: The heart of the redox hub. Plant Physiology. 2011;**155**:2-18

[51] Foyer CH, Noctor G. Redox sensing and signalling associated with reactive oxygen in chloroplasts, peroxisomes and mitochondria. Physiologia Plantarum. 2003;**119**:355-364

[52] Blokhina O, Virolainen E, Fagerstedt KV. Antioxidants, oxidative damage and oxygen deprivation stress: A review. Annals of Botany. 2003;**91**:179-194

[53] Dumanović J, Nepovimova E, Natić M, Kuča K, Jaćević V. The significance of reactive oxygen species and antioxidant defense system in plants: A concise overview. Frontiers in Plant Science. 2021;**11**:552969

[54] Cao Q, Lv W, Jiang H, Chen X, Wang X, Wang Y. Genome-wide identification of glutathione S-transferase gene family members in tea plant (*Camellia sinensis*) and their response to environmental stress. International Journal of Biological Macromolecules. 2022;**205**:749-760

[55] Kaur A, Tyagi S, Singh K, Upadhyay SK. Exploration of glutathione reductase for abiotic stress response in bread wheat (*Triticum aestivum* L.). Plant Cell Reports. 2022;**41**(3):639-654

[56] Yin X, Bai YL, Ye T, Yu M, Wu Y, Feng YQ. Cinnamoyl coA: NADP oxidoreductase-like 1 regulates abscisic acid response by modulating phaseic acid homeostasis in *Arabidopsis thaliana*. Journal of Experimental Botany. 2022;**73**(3):860-872

[57] Lee SC, Choi DS, Hwang IS, Hwang BK. The pepper oxidoreductase CaOXR1 interacts with the transcription factor CaRAV1 and is required for salt and osmotic stress tolerance. Plant Molecular Biology. 2010;**73**(4):409-424

[58] Corpas FJ, Palma JM, Sandalio LM, Valderrama R, Barroso JB, Luis A. Peroxisomal xanthine oxidoreductase: Characterization of the enzyme from pea (*Pisum sativum* L.) leaves. Journal of Plant Physiology. 2008;**165**(13):1319-1330

[59] Gutsche N, Thurow C, Zachgo S, Gatz C. Plant-specific CC-type glutaredoxins: Functions in developmental processes and stress responses. Biological Chemistry. 2015;**396**(5):495-509

[60] Brito DS, Quinhones CG, Neri-Silva R, Heinemann B, Schertl P, Cavalcanti JHF, et al. The role of the electron-transfer flavoprotein: Ubiquinone oxidoreductase following carbohydrate starvation in Arabidopsis cell cultures. Plant Cell Reports. 2022;**41**(2):431-446

[61] da Fonseca-Pereira P, Pham PA, Cavalcanti JHF, Omena-Garcia RP, Barros JA, Rosado-Souza L, et al. The Arabidopsis electron-transfer flavoprotein: Ubiquinone oxidoreductase is required during normal seed development and germination. The Plant Journal. 2022;**109**(1):196-214

[62] Onda Y, Okino T. Thiol-disulfide oxidoreductase PDI1; 1 regulates actin structures in Oryza sativa root cells. FEBS letters. 2022;**596**(23):3015-3023

Chapter 4

The Good and the Bad: The Bifunctional Enzyme Xanthine Oxidoreductase in the Production of Reactive Oxygen Species

Brandon Charles Seychell, Marita Vella, Gary James Hunter and Thérèse Hunter

Abstract

Xanthine oxidoreductase (XOR) is a molybdoflavin enzyme which occurs in two forms; the reduced form known as xanthine dehydrogenase (XDH, EC 1.17.1.4) and the oxidised form known as xanthine oxidase (XO, EC 1.17.3.2). In humans, it is a 293 kDa homodimer which catalyses consecutive hydroxylation steps of purine degradation. The oxidised form of the enzyme produces hydrogen peroxide and superoxide ($O_2^{\bullet -}$), both of which are reactive oxygen species (ROS) that can interact with several biomolecules producing adverse reactions. XOR can also produce nitric oxide, a cardiovascular protective molecule. Overproduction of nitric oxide results in the formation of the highly reactive peroxynitrite radical. XOR-produced ROS may provide protection against infection, while at the same time can also lead to inflammation, oncogenesis, brain injury and stroke. XOR is also involved in tumour lysis syndrome in chemotherapy patients as well in ischaemia-reperfusion injury, increasing the levels of ROS in the body. Consequently, the presence of XOR in blood can be used as a biomarker for a number of conditions including oxidative stress and cardiovascular disease.

Keywords: xanthine oxidoreductase, reactive oxygen species, superoxide, cardiovascular disease, oxidative stress

1. Introduction

Xanthine Oxidoreductase (XOR) is ubiquitous amongst organisms being found in both prokaryotes and eukaryotes. It is a complex flavoenzyme and one of a few that require a molybdenum cofactor (Moco), in the form of molybdopterin (MPT). As well as a requirement for flavin adenine dinucleotide (FAD) it also contains two iron–sulphur ([2Fe-2S]) clusters. XOR exists in two forms: most organisms have only the dehydrogenase, XDH (EC 1.17.1.4), while mammals also produce the oxidase, XO (EC 1.17.3.2). The XDH to XO conversion has been studied [1, 2]. It has been demonstrated that conversion could be accomplished in two ways; either through

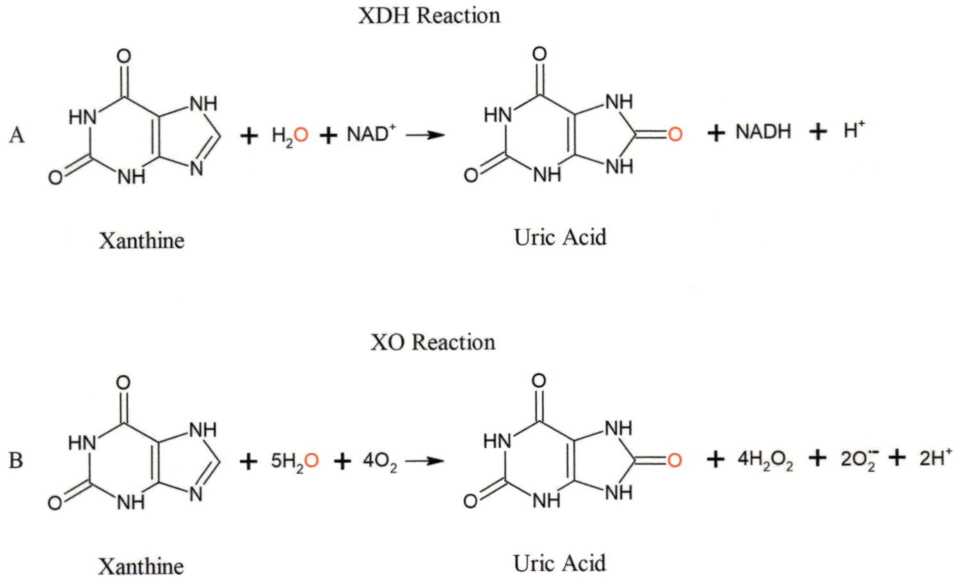

Figure 1.
Xanthine oxidoreductase (XOR) catalysed reactions. A. Xanthine dehydrogenase (XDH) reaction with NAD+ as electron acceptor. B. Xanthine oxidase (XO) reaction with molecular oxygen as electron acceptor. Note that the majority of XOR enzymes are found in the XDH form while mammalian enzymes may be converted from XDH to XO (see text for details).

disulphide formation of specific cysteine residues or by limited proteolysis. The major reactions catalysed by the dehydrogenase (XDH), that of oxidising hypoxanthine to xanthine and then xanthine to uric acid (**Figure 1A**), are important last steps in the degradation of purines (and therefore in the degradation of DNA and RNA). The oxidase (XO) on the other hand, can generate reactive oxygen species (ROS) which are involved in the inflammatory defensive response (**Figure 1B**) [3].

The enzyme also has nitrite reductase activity and is therefore important in the production of nitric oxide and involved in the physiological control of blood pressure. In this review we discuss the structure, catalytic mechanism, and tissue localisation of human XOR and its role in health and disease.

2. Mammalian xanthine oxidoreductase

Mammalian XOR exists as a homodimer with a molecular weight of about 290 kDa (**Figure 2**). XOR can be found in both XO and XDH forms and has a wide distribution in mammalian tissues, with the highest amounts observed in the intestines and liver. XOR is present in milk and has in fact, been detected in mammary gland epithelial and capillary endothelial cells. Intracellularly, XDH is the predominant species whilst XO is present in the gastrointestinal lumen and body fluids. Inactive forms of XOR do occur in mammalian tissues and usually lack either molybdenum or sulphur in the Moco cofactor. It was estimated that up to 60% of XOR found in bovine milk may be in an inactive state [4]. Demolybdo XOR, lacks the molybdenum (Mo) and possibly the whole MPT cofactor, while desulfo XOR, has a molybdenum-bound sulphur atom replaced by oxygen in the Moco. Due to the presence of these inactive forms, immunodetection of XOR may not correlate well with level of XOR activity. There

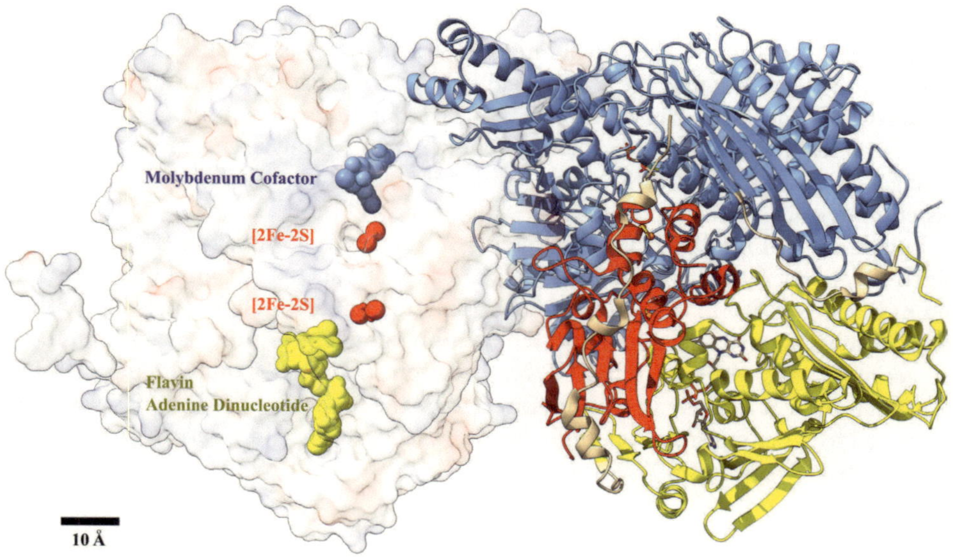

Figure 2.
*Crystal structure of bovine xanthine oxidoreductase, (PDB 3UNC). The two subunits of the biologically active dimeric enzyme are shown as different representations. On the right-hand side is a cartoon of one monomer; blue represents the Moco domain (including the C terminus), yellow represents the FAD domain and red represent the iron–Sulphur domain (and includes the N-terminus). Cream represents the link portions and cofactors are also shown in stick form. On the left hand side the monomer is shown as a surface coloured by electrostatic potential with cofactors as spheres coloured as their domains in the cartoon and labelled accordingly. Note that the cofactors in each monomer are orientated similarly from top to bottom, which is also the direction of the flow of electrons during the catalytic reaction (**Figure 3**).*

is considerable homology between XOR and aldehyde oxidases, exemplified by the observed cross-reactivity of XOR antibodies, further complicating the analysis by immunodetection [5, 6].

3. Human xanthine oxidoreductase

Human XOR activity has been found to be low in organs other than the liver and the intestines [5, 6]. XOR was shown to be a source of reactive oxygen species (ROS) and reactive nitrogen species (RNS) in vasculature and digestive tract. In an inflammatory response, it has been shown that XOR protects against infection through the generation of ROS and RNS, and so was designated as an innate immunity agent [7]. Additionally, it is deduced that XOR provides antimicrobial defence in the gastrointestinal tract where it can concentrate on the epithelial cells' surface [8]. Milk XOR has bactericidal characteristics and provides immunity in the new-born gut, where other immune responses have not yet formed [9]. In the liver, XOR may catalyse the oxidation of numerous metabolites such as purines, pyrimidines, heterocyclic compounds, and aldehydes, as well as other xenobiotics such as anti-cancer drugs. As a result, XOR contributes to liver detoxification [10].

At normal physiological conditions, XOR is present in very low amounts in serum [11]. Due to the presence of serum proteases, circulating XOR is virtually all in XO form [12]. During certain disease states, the levels of blood XO increase. For example, during viral hepatitis, the level of blood XO is 1000 times higher than the controls [13]. Other disease states were blood XO was found to be elevated include artherosclerosis, mixed connective tissue disease and scleroderma [14, 15]. Moreover, the level of circulating XO has been reported to also increase during ischaemia-reperfusion injury induced by liver transplantation or aortic cross-clamp procedures amongst others [16, 17]. Circulating XO can also bind to glycoslaminoglycans on the vascular endothelial cells' surface and, if concentrated enough, can cause oxidative damage to organs that are far from the original site of XO release/damage and/or have low content of XOR during normal physiological conditions [18].

Purification of human XOR from liver tissue and breast milk, and (recombinantly) from bacteria has been described [19–21]. The activity of pure human liver and milk XO was found to be just 5% that of bovine milk, which is consistent with the molybdenum content measurement, where human milk XOR only has 4% molybdenum content, while bovine milk XOR has between 70 and 60% molybdenum content [4, 20]. Human XOR activity was observed to be the highest in the liver. However, as compared to other mammals, namely *Bos taurus*, *Ovis aries* and *Capra aegagrus,* the specific activity of human XOR is low [4, 22, 23].

4. Structure and catalysis

Due to the evolutionary loss of uricase in humans and higher primates, XOR serves as the ultimate catalyst in purine degradation, catalysing the last two hydroxylation reactions by first converting hypoxanthine to xanthine, and then xanthine to uric acid [24].

The first XDH and XO crystal structures were elucidated by Enroth and co-workers using protein isolated from bovine milk [25]. The crystal structure of bovine XOR shows that it is organised as a homodimer with three redox-centered domains per monomer. Each subunit consists of a polypeptide chain of 1332 amino acids of 146.8 kDa and contains two iron sulphur clusters (each [2Fe-2S]), one Moco and one flavin adenine dinucleotide (FAD) coenzyme (**Figure 2**).

The 20 kDa N-terminal domain that houses the two [2Fe-2S] is connected via a linker peptide to a central 40 kDa FAD domain. This is, in turn, connected by another linker to the larger 85 kDa C-terminal molybdopterin domain. The molybdopterin lies close to the interfaces of both the Fe-S and the FAD domains. The two hydroxylation reactions that convert hypoxanthine to xanthine, and xanthine to urate both occur at the Moco active site of each monomer.

During catalysis, electrons flow sequentially through the redox centers of the three domains in a nearly linear manner commencing at the Moco domain where the oxidative half-reaction takes place, to the two [2Fe-2S] domains, and finally, the FAD domain, where the reductive half-reaction takes place. Here NAD^+ is reduced to NADH (**Figure 3**). As the two Mo centers in the homodimer are 52 Å apart, no electron transfer is possible between the two monomers (**Figure 2**) [25].

The active site Glu-1261 (bovine numbering; sequence accession number P80457) initiates the reaction by abstraction of a proton from a hydroxide bound to Mo at the Moco site (**Figure 4**). This results in a deprotonated ion, Mo-O^-, that is nucleophilic and attacks carbon-8 of the xanthine substrate. This causes a hydride shift from the substrate onto the Moco sulphur, changing the valency of the Mo from (VI) to (IV). The subsequent reactions depend on the reaction conditions and substrate. If electron transfer from the Mo center to the sulphur precedes the product displacement, Path A (**Figure 4**), occurs. This path occurs at high pH and substrate concentrations, and with slow substrates, such as 2-hydroxy-6-methylpurine [26]. Under these conditions,

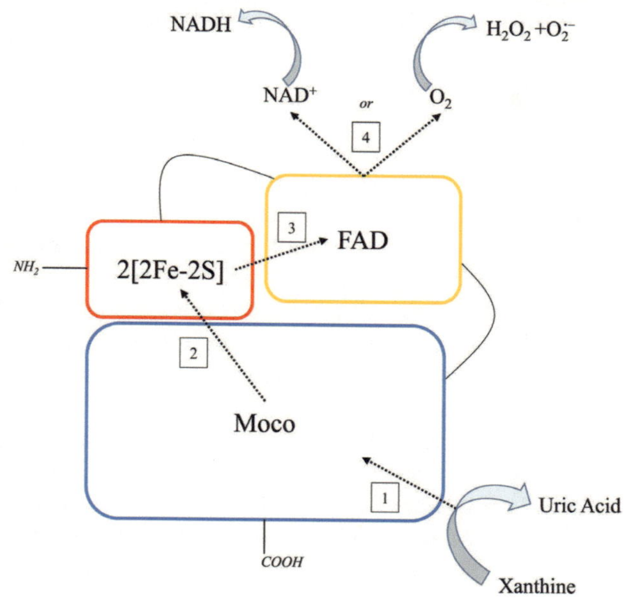

Figure 3.
Electron transfer in XOR. Arrows represent electron transfer. Xanthine donates its electrons at the Moco site (1) and oxidises to form uric acid in the process. The electrons propagate through the 2[2Fe-2S] clusters (2) to the FAD (3) and are then ultimately accepted by either NAD+ (XDH reaction) or oxygen (XO reaction), depending on whether the enzyme is reduced or oxidised respectively (4). Superoxide production occurs when oxygen is the electron acceptor.

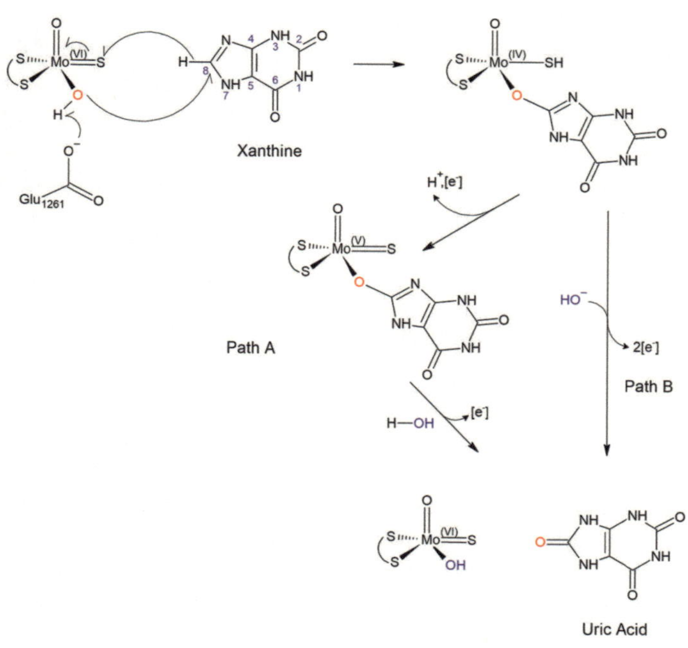

Figure 4.
Reaction mechanism of XOR with xanthine as a substrate. Glu-1261 abstracts a hydrogen from the hydroxyl group coordinated with Mo, resulting in a cascade of reactions, ending with the first intermediate (top right). This intermediate, can go through one of two alternative paths. Which pathway takes place is dependent on the reaction conditions; Path A occurs at high pH or high substrate concentrations, and slow substrates, whereas path B occurs at a pH lower than 8.3.

a Mo(V) species will form which displaces the product and forms the Mo(VI) species via hydrolysis. Under normal physiological conditions, however, the Mo(V) intermediate is skipped and product dissociation occurs first (Path B, **Figure 4**) [26, 27].

The XDH and XO forms of the mammalian XOR are mechanistically distinct as XDH preferentially uses NAD^+ as the final electron acceptor at the FAD site, forming NADH, while oxygen is the final electron acceptor in XO which produces hydrogen peroxide and superoxide radical (**Figure 3**). These XO-generated ROS may serve a physiological purpose but may also have pathological consequences. In mammals XDH can be converted to XO either by the reversible oxidation of cysteine residues to form a disulfide bond or by irreversible limited proteolysis [28, 29]. The formation of XO has three effects: hindrance of the entry of NAD^+ to the FAD active site, the formation of an access channel to this active site and an increase in the redox potential that supports the generation of superoxide [30].

As XO is the principle source of XOR-produced ROS, this conversion process is of particular interest. The mechanism of XDH-XO conversion was elucidated by comparing *Bos taurus* XOR, *Rattus norvegicus* XOR (both of which can be found in both XDH and XO forms) and *Gallus gallus* XOR (can be found only as the XDH form). On comparing the *R. norvegicus* XOR with the *G. gallus* XOR, Nishino et al. showed that two pairs of cysteine residues were responsible for formation of disulphide bridges during this conversion: Cys535 with Cys992 and Cys1316 with Cys1324 (*R. norvegicus* numbering) [31].

In the bovine XOR crystal structure, Cys992 is positioned on the molecular surface, while Cys535 is on a linker peptide (Lys532 - Ser589) between the Moco and FAD. Cys535-Cys992 disulphide bridge has a far-reaching conformational effect on the Gln422 - Lys432 C-terminal loop that would otherwise structurally contribute the FAD active site in XOR. The displaced loop partly covers the FAD active site in XO, thus blocking NAD^+ from entering and acting as the electron acceptor. Furthermore, this disturbance alters the electrostatic environment surrounding the FAD, making the semiquinone less stable and more reactive to oxygen. As a result, oxygen becomes the primary electron acceptor in XO [25]. NAD^+ reactivity in XO is further decreased due to the oxidation and disulfide bond formation at Cys1316 and Cys1324. Upon oxidation, the XOR C-terminus is obstructed from interacting with the NAD^+ binding cavity. This interaction was shown to be critical for NAD^+ binding in the bovine structure [31]. Similarly, irreversible proteolysis near Leu 215 and Leu 569 causes considerable structural disturbance of the Gln422 - Lys432 C-terminal loop that interferes with the reaction at the FAD (**Figure 5**).

Apart from its canonical activity, XOR can also act as a nitrite reductase by reducing NO_2^- to $NO^•$ at the Moco site using either xanthine or NADH as the electron donors [32, 33]. Li et al. noted that the site of electron donation differs depending on the electron donor involved in the reaction [34]. While NADH donates its electrons to the FAD site, xanthine donates its electrons directly to the Moco site. Under normal physiological conditions, the preference for xanthine as an electron acceptor to produce uric acid inhibits nitric oxide ($NO^•$) formation. As such, it is suggested that NADH acts as the primary electron donor for $NO^•$ production, indicating that XDH, not XO, is the preferred catalyst under hypoxic conditions [35]. The presence of oxygen results in lower production of $NO^•$ due to oxidation of the Moco site, while NO is formed under hypoxic conditions. Additionally, $NO^•$ formation is an acid catalysed reaction and only occurs under acidic pH conditions such as those observed during inflammation or intracellular acidosis in ischaemic conditions where pH falls to 6.0 or below [36].

5. XOR-generated ROS/RNS

Although chronic and elevated levels of ROS/RNS are culpable of extensive cellular and molecular damage and identified as primary suspects in pathogenicity, when within threshold levels, reactive species do play vital roles in cellular signalling, vasodilation, neutrophil burst and neurotransmission for example [37, 38]. There is, therefore, a fine line that separates the benefits exerted by these reactive species from the harm they may cause that is in turn influenced by the activity of circulating XO.

Expression of XOR in arterial endothelial cells increases in response to hypoxia and inflammatory cytokines stimuli [39]. Once secreted, the circulating XO can bind to glycosaminoglycans (GAGs) on the surface of endothelial cells, leading to localised amplification of XO activity [40, 41]. XOR is converted to XO especially during inflammation and ischaemia-reperfusion injury. Ischaemia is a condition characterised by limited oxygen supply, while reperfusion refers to the restoration of blood flow and oxygen to tissues. During ischaemia, a rise in calcium levels activates calpain protease, which irreversibly converts XDH to XO through limited proteolysis [42]. Furthermore, under these conditions, levels of hypoxanthine also increase due to the release of ATP into the extracellular space [43]. Although XO is not the only molecule that causes vascular damage due to oxidative stress, it is of particular interest because it is a known druggable target.

Levels of ROS also continue to rise during the reperfusion stage as XDH is oxidatively converted to XO. The level of oxygen tension influences the type of ROS that forms, with superoxide being the predominant ROS species at higher oxygen tension while more hydrogen peroxide is produced as hypoxia worsens. Increases in hydrogen peroxide have far-reaching physiological effects since it is a significant regulator of signalling pathways [44, 45].

XDH too may contribute to ROS levels. Although oxygen is not a preferred substrate of XDH, it may still be reduced by XDH under conditions of ischaemia/hypoxia when NAD^+ levels are characteristically low due to a decrease in mitochondrial oxidation of NADH (**Figure 6**) [40].

Superoxide may cause further damage by reacting with hydrogen peroxide or nitric oxide. Even though H_2O_2 is not a very potent ROS, in the presence of iron or copper catalysts it may react with $O_2^{\bullet-}$ to produce the more reactive hydroxyl radicals ($OH^\bullet$) via the Fenton reaction [46]. Superoxide may also react with nitric oxide ($NO^\bullet$), which produces peroxynitrite ($ONOO^-$). The latter is a very strong oxidising agent that is capable of damaging biomolecules, including aconitase, Hsp90, and mitochondrial membrane components [47–49].

Polyunsaturated fatty acids are highly susceptible to lipid peroxidation by ROS that attack polyunsaturated lipids to produce highly reactive lipid peroxy radicals and 4-hydroxyononenal [50]. These reactive species can then react with DNA and membrane proteins, leading to changes in the fluidity and permeability of the lipid bilayer [51]. ROS have been linked to neurodegenerative disorders and acute neurological diseases, including strokes and epilepsy [52, 53]. This may be due to the production of $OH^\bullet$ radicals via the Fenton reaction, as the brain has a high concentration of iron and copper. Research suggests that XO contributes to brain injury and stroke [53].

Reactive nitrogen species (RNS) are a group of nitrogen-based compounds that function as highly reactive free radicals. This includes nitrogen dioxide, dinitrogen trioxide, and peroxynitrite ($ONOO^-$). The $NO^\bullet$ and $O_2^{\bullet-}$ that may both be generated by XOR, can react with each other to produce $ONOO^-$ [32, 54]. This leads to the rapid drop in the available amount of $NO^\bullet$. Since $NO^\bullet$ is a molecule that protects

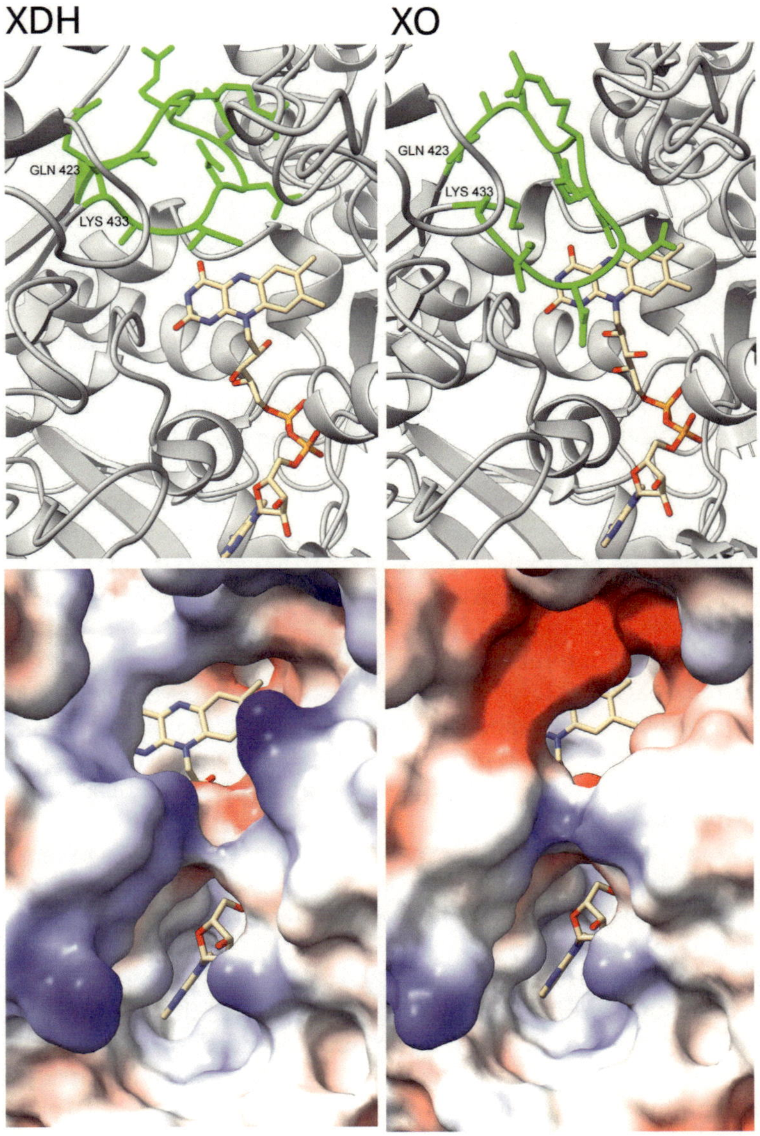

Figure 5.
The electrostatic environment around the XOR FAD domain in XDH (left panels) and XO (right panels). Top: The Gln423-Lys433 loop (green) with respect to the FAD where the loop covers the FAD's quinone moiety in XO (right panel). Bottom: Electrostatic surface around the FAD where blue and red regions represent electropositive and electronegative regions respectively. XDH contains more electropositive areas surrounding the FAD and larger active site pores, allowing both oxygen and NAD^+ to enter the active site. XO contains more electronegative areas around FAD and narrower pores, preventing the NAD^+ approach.

the cardiovascular system, its deficiency may contribute to various cardiovascular diseases such as atherosclerosis, hypertension, and coronary artery disease [55].

The production of $ONOO^-$ is linked to the location where $O_2^{\bullet -}$ is produced. This is because $O_2^{\bullet -}$ has a shorter half-life and is less able to diffuse than $NO^{\bullet}$. Although $ONOO^-$ has a very brief half-life of approximately 10 ms under normal pH conditions, it can still permeate cell membranes and interact with molecules in neighbouring cells [56]. In fact, the impact of $ONOO^-$ can be felt up to a distance of 20 μm from its origin [49].

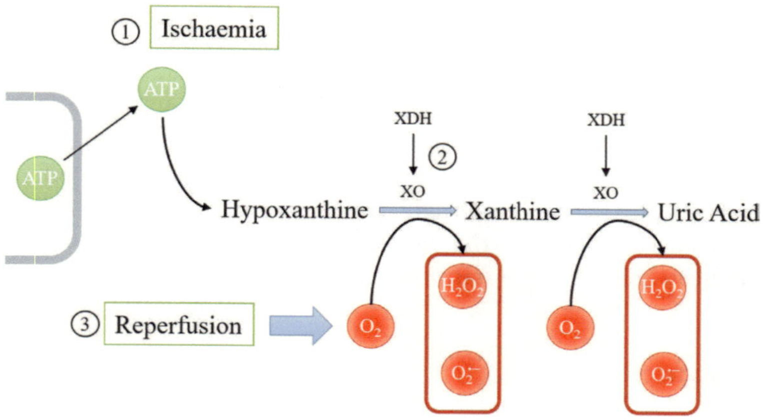

Figure 6.
Ischaemia-reperfusion effects on XOR action. 1. During ischaemia ATP is released into the extracellular space where it is catabolised to adenosine by membrane bound enzymes. Adenosine is further catabolised to hypoxanthine. 2. XDH is converted to XO due to limited proteolysis during ischaemia or oxidation during reperfusion. 3. Oxygen supply is back during reperfusion which can act as an electron acceptor producing reactive oxygen species marked in red.

The highly reactive and oxidative $ONOO^-$ anion may damage cellular structures such as mitochondria by inactivating manganese superoxide dismutase [57]. $ONOO^-$ also oxidises cysteine and nitrates tyrosine residues in membrane channels, including calcium pumps, which impairs cellular ionic balance [58]. Additionally, $ONOO^-$ induces DNA double-strand breaks that can cause cell death and reduces the amount of antioxidants like cysteine, which consequently leads to an increase in free-radical damage [59, 60].

6. Uric acid

Impaired kidney urate excretion is a common cause of hyperuricemia as are excessive alcohol consumption, high protein diet and chemotherapy. The underlying inflammatory state that is maintained by chronic hyperuricemia (6.8 mg/dl) exacerbates the symptoms of metabolic syndrome and is associated with type 2 diabetes, hypertension and kidney inflammatory disease [61–63]. This urate-induced inflammatory response is also a risk factor for cardiovascular disease including stroke and heart failure.

The association of low, but not high, urate levels with neurological disorders such as multiple sclerosis, Parkinson's and Huntington is curious as in these pathologies, hyperuricemia was in fact, deemed responsible of slower progression of the disease [64, 65]. It is hypothesised that the antioxidant behaviour of urate as a scavenger of superoxide radicals is neuroprotective [66, 67].

7. XOR in cardiovascular disease and hypertension

Evidence shows that elevated levels of XOR and its associated oxidative stress are risk factors of cardiovascular disease, causing endothelial damage, hypertension and affecting vascular tone. In particular, the protein damage due to the nitration of tyrosine residues by peroxynitrite has been implicated in the pathophysiology of cardiovascular disease [68]. Whether hyperuricaemia contributed to cardiovascular disease, especially hypertension, has been a controversial topic. Evidence suggests that the accumulation of intracellular urate may be associated with hypertension due to an increase in oxidative stress and inflammation [69]. while extracellular urate may cause calcification of blood vessels [69, 70]. Recent studies by Yoshida et al. and Furuhashi et al. conclude that serum levels of XOR are positively and independently associated with hypertension via the oxidative stress it forms [71, 72]. A number of polymorphisms in the XDH gene that increase enzyme activity such as G172R, A932T, N1109T have been linked to hypertension and related morbidities [73].

8. XOR and cancer

Although the level of XOR expression varies in different cancer types, dysregulated XOR expression and alterations in activity have been observed in various cancers and appear to be closely associated with clinical outcomes [74, 75]. Due to the biochemical complexity of XOR, it is not easy to predict or determine its involvement in cancer development and progression nor whether XOR inhibitors may be beneficial in prevention. The activity of XOR is greatly influenced by its microenvironment, its isoform state, cofactor availability, its broad substrate specificity and its multiple catalytic properties as a xanthine dehydrogenase, a xanthine oxidase, a NADH oxidase and nitrite/nitrate reductase. There is also the issue of which and how much product is formed as some may have pro- while others may have anti-tumorigenic potential. The final products of XDH are uric acid and NADH, although superoxide and hydrogen peroxide may be formed when oxygen is limited. XO generates uric acid, superoxide, hydrogen peroxide and when in the presence of nitrites/nitrates, nitric oxide. To complicate matters further, not only are these ROS capable of directly causing cellular damage, but both also serve as signalling molecules that affect tissue physiology [76–78].

The irreversible degradation of purines by XOR limits the formation of nucleotides by the salvage pathway which in turn limits cellular proliferation. The expression of XOR responds to various signals that include oxygen tension and inflammatory cytokines [79]. In breast, ovarian, gastric, lung and colorectal cancer tissue, significant down-regulation of XOR expression was observed in advanced, aggressive cancer stages. The loss of XOR expression is suggested to be an independent predictor for poor patient outcomes [80–84]. In fact, the situation is different during early stages of oncogenic transformation.

Xu et al. report that alternol, a natural compound, can effectively trigger a significant response of ROS and cell death through activating XO. This activation occurs when alternol interacts with the catalytic molybdenum-binding domain of XOR, without interfering with the FAD cofactor. According to computational analysis, alternol interacts with the catalytic domain of the XOR protein and enhances its oxidative activity, leading to an increase in XOR activity and proteolytic processing in prostate cancer cells, but not in benign cells. This might provide novel insights on how to treat cancer without harming other tissues [85].

9. XOR inhibitors

XOR inhibitors (XOI) that reduce elevated serum urate levels in the treatment of gout, also curb oxidative stress and increase adenosine triphosphate levels. These drugs are prescribed when the concentration of urate in the bloodstream is 6.8 mg/dl or more [86]. Allopurinol, a purine analogue, has been on the market since 1956 and is the most prescribed XOI in clinical settings and is administered for the treatment of gout and the prevention of tumour lysis syndrome as a consequence of cancer chemotherapy. The rapid increase in uric acid levels characteristic of tumour lysis syndrome may precipitate death. Allopurinol is effective and relatively inexpensive, being a non-specific competitive and suicide inhibitor of XOR. XOR transforms allopurinol into oxypurinol, which binds to the Moco site's Mo(IV) thereby inhibiting catalysis of the hypoxanthine/xanthine substrates [87]. Allopurinol may cause various side effects such as nephropathy, hepatitis, Stevens-Johnson syndrome, and allopurinol hypersensitivity syndrome. A second XOI is febuxostat. This is a non-purine specific inhibitor that binds tightly to the Moco site of both forms of XOR and is more effective than allopurinol at lowering serum urate. The common side effects of febuxostat include diarrhoea, nausea, and elevated levels of liver enzymes. The main disadvantage of febuxostat is its cost, and therefore, it is considered a second choice after allopurinol.

The third XOI is topiroxostat. This a non-purine analogue that exhibits mechanistic properties of both allopurinol and febuxostat. It initially binds in a competitive manner instead of the substrate to the XOR active site after which it binds covalently with the molybdenum and blocks the Moco site. The dissociation of topiroxostat from the Moco site has a prolonged half-life of 20 hours [88, 89]. Although allopurinol and febuxostat have been approved for use in the United States, European Union, and Japan, topiroxostat is currently only approved for use in Japan.

10. Conclusion

Despite being an evolutionally ancient and highly conserved enzyme that has been studied since the early 1900s, xanthine oxidoreductase does not fail to surprise [90]. What has clearly emerged over these years is that XOR is a biomolecule with many options. As an enzyme, XOR is multifunctional protein with a broad substrate specificity. This not only permits it to catalyse the final two consecutive hydroxylation steps in the degradation pathway of purines but also to detoxify many other endogenous and exogenous molecules. The XOR protein itself can exist in two forms, XDH which may be converted post-translationally, in a reversible or irreversible manner, to XO in response to the microenvironment, oxygen tension and various other signals. This makes it versatile and capable of adapting rapidly to physiological signals. Even expression of its gene is tightly regulated by physiological stimuli such as inflammatory cytokines. Although both isoforms conduct the same reactions at the Moco site to produce uric acid, XO is sufficiently structurally distinct at the FAD domain to only permit the binding of oxygen as the final electron acceptor at this active site. This results in the generation of superoxide and hydrogen peroxide reactive species rather than NADH. These ROS in turn, exert diverse physiological effects that include amongst others anti-microbial activity, wound healing, proliferation, cellular ageing. The versatility of XOR catalysis includes a nitrite reductase activity that results in the formation of nitric oxide that is an essential signalling molecule in vascular physiology.

The complicated enzymatic profile of XOR does benefit many physiological systems but when dysregulated may result in pathological outcomes. The levels of ROS may damage endothelial cells, may form highly oxidative peroxynitrite and hydroxyl radicals, and may disrupt cellular pathways, cause inflammation and all its ensuing downstream effects. Changes in expression and activity are influenced by external factors including diet, alcohol and drugs while clinically relevant polymorphisms have been associated with hypertension.

The close relationship between XOR activity and certain pathophysiologies including cancer, metabolic syndrome and hypertension highlights the potential of XOR as a biomarker and as a therapeutic target.

Acknowledgements

This chapter was funded by the project FABXS financed by the Malta Council for Science & Technology, for and on behalf of the Foundation for Science and Technology, through the FUSION: R&I Technology Development Programme and by the project Crystallisation of Protein Targets financed by the Malta Council for Science & Technology, for and on behalf of the Foundation for Science and Technology, through the Internationalisation Partnership Awards Scheme Plus, IPAS-2022-009.

Author details

Brandon Charles Seychell, Marita Vella, Gary James Hunter and Thérèse Hunter*
Department of Physiology and Biochemistry, University of Malta, Msida, Malta

*Address all correspondence to: therese.hunter@um.edu.mt

References

[1] Rajagopalan KV, Handler P. Purification and properties of chicken liver xanthine dehydrogenase. The Journal of Biological Chemistry. 1967;**242**:4097-4107

[2] Kanda M, Brady FO, Rajagopalan KV, Handler P. Studies on the dissociation of flavin adenine dinucleotide from metalloflavoproteins. The Journal of Biological Chemistry. 1972;**247**:765-770

[3] Bortolotti M, Polito L, Battelli MG, Bolognesi A. Xanthine oxidoreductase: One enzyme for multiple physiological tasks. Redox Biology. 2021;**41**:101882. DOI: 10.1016/j.redox.2021.101882

[4] Ventom AM, Deistung J, Bray RC. The isolation of demolybdo xanthine oxidase from bovine milk. The Biochemical Journal. 1988. DOI: 10.1042/bj2550949

[5] Jarasch ED, Grund C, Bruder G, Heid HW, Keenan TW, Franke WW. Localization of xanthine oxidase in mammary-gland epithelium and capillary endothelium. Cell. 1981;**25**:67-82. DOI: 10.1016/0092-8674(81)90232-4

[6] Harrison R. Structure and function of xanthine oxidoreductase: Where are we now? Free Radical Biology & Medicine. 2002;**33**:774-797. DOI: 10.1002/ejic.200600087

[7] Martin HM, Hancock JT, Salisbury V, Harrison R. Role of xanthine oxidoreductase as an antimicrobial agent. Infection and Immunity. 2004;**72**:4933-4939. DOI: 10.1128/IAI.72.9.4933-4939.2004

[8] Frederiks WM, Vreeling-Sindelarova H. Ultrastructural localization of xanthine oxidoreductase activity in isolated rat liver cells. Acta Histochemica. 2002;**104**:29-37. DOI: 10.1078/0065-1281-00629

[9] Harrison R. Milk xanthine oxidase: Properties and physiological roles. International Dairy Journal. 2006;**16**:546-554. DOI: 10.1016/j.idairyj.2005.08.016

[10] Kathriarachchi K, Leus M, Everett DW. Oxidation of aldehydes by xanthine oxidase located on the surface of emulsions and washed milk fat globules. International Dairy Journal. 2014;**37**:117-126. DOI: 10.1016/J.IDAIRYJ.2014.03.007

[11] Shamma'a MH, Nasrallah SM, Al-Khalidi UAS. Serum xanthine oxidase – An experience with 2000 patients. The American Journal of Digestive Diseases. 1973;**18**:15-22. DOI: 10.1007/BF01072233

[12] Kooij A, Schiller HJ, Schijns M, Van Noorden CJF, Frederiks WM. Conversion of xanthine dehydrogenase into xanthine oxidase in rat liver and plasma at the onset of reperfusion after ischemia. Hepatology. 1994;**19**:1488-1495. DOI: 10.1002/hep.1840190626

[13] Murase T, Oka M, Nampei M, Miyachi A, Nakamura T. A highly sensitive assay for xanthine oxidoreductase activity using a combination of [$^{13}C_2$, $^{15}N_2$]xanthine and liquid chromatography/triple quadrupole mass spectrometry. Journal of Labelled Compounds and Radiopharmaceuticals. 2016;**59**:214-220. DOI: 10.1002/jlcr.3390

[14] Mohàcsi A, Kozlovszky B, Kiss I, Seres I, Fülöp T. Neutrophils obtained from obliterative atherosclerotic patients exhibit enbanced resting respiratory burst and increased degranulation in

response to various stimuli. Biochimica et Biophysica Acta – Molecular Basis of Disease. 1996;**1316**:210-216. DOI: 10.1016/0925-4439(96)00027-0

[15] Miesel R, Zuber M. Elevated levels of xanthine oxidase in serum of patients with inflammatory and autoimmune rheumatic diseases. Inflammation. 1993;**17**:551-561. DOI: 10.1007/BF00914193

[16] Battelli MG, Ravaioli M, Musiani S, Scicchitano V, Grazi GL, Bolognesi A. Increased serum level of xanthine oxidoreductase in liver transplanted patients. Journal of Biological Regulators and Homeostatic Agents. 2011;**25**:77-84

[17] Sidhartha T, Gelman S, Wheat JK, Parks DA. Circulating xanthine oxidase in human ischemia reperfusion. Southern Medical Journal. 1995;**88**:479-482. DOI: 10.1097/00007611-199504000-00021

[18] Adachi T, Fukushima T, Usami Y, Hirano K. Binding of human xanthine oxidase to sulphated glycosaminoglycans on the endothelial-cell surface. The Biochemical Journal. 1993;**289**:523-527. DOI: 10.1042/bj2890523

[19] Spector T, Hall WW, Krenitsky TA. Human and bovine xanthine oxidases. Inhibition studies with oxipurinol. Biochemical and Pharmacology. 1986;**35**:3109-3114. DOI: 10.1016/0006-2952(86)90394-1

[20] Godber BLJ, Schwarz G, Mendel RR, Lowe DJ, Bray RC, Eisenthal R, et al. Molecular characterization of human xanthine oxidoreductase: The enzyme is grossly deficient in molybdenum and substantially deficient in iron-sulphur centres. The Biochemical Journal. 2005;**388**:501-508. DOI: 10.1042/BJ20041984

[21] Yamaguchi Y, Matsumura T, Ichida K, Okamoto K, Nishino T. Human xanthine oxidase changes its substrate specificity to aldehyde oxidase type upon mutation of amino acid residues in the active site: Roles of active site residues in binding and activation of purine substrate. Journal of Biochemistry. 2007;**141**:513-524. DOI: 10.1093/jb/mvm053

[22] Atmani D, Benboubetra M, Harrison R. Goats' milk xanthine oxidoreductase is grossly deficient in molybdenum. The Journal of Dairy Research. 2004;**71**:7-13. DOI: 10.1017/S0022029903006514

[23] Benboubetra M, Baghiani A, Atmani D, Harrison R. Physicochemical and kinetic properties of purified Sheep's Milk xanthine oxidoreductase. Journal of Dairy Science. 2004;**87**:1580-1584. DOI: 10.3168/jds.S0022-0302(04)73311-1

[24] Oda M, Satta Y, Takenaka O, Takahata N. Loss of urate oxidase activity in hominoids and its evolutionary implications. Molecular Biology and Evolution. 2002;**19**:640-645. DOI: 10.1093/oxfordjournals.molbev.a004123

[25] Enroth C, Eger BT, Okamoto K, Nishino T, Pai EF. Crystal structures of bovine milk xanthine dehydrogenase and xanthine oxidase: Structure-based mechanism of conversion. Proceedings of the National Academy of Sciences. 2000;**97**:10723-10728. DOI: 10.1073/pnas.97.20.10723

[26] Xia M, Dempski R, Hille R. The reductive half-reaction of xanthine oxidase. Reaction with aldehyde substrates and identification of the catalytically labile oxygen. The Journal of Biological Chemistry. 1999;**274**:3323-3330. DOI: 10.1074/jbc.274.6.3323

[27] Hille R, Hall J, Basu P. The mononuclear molybdenum enzymes. Chemical Reviews. 2014;**114**:3963-4038. DOI: 10.1021/cr400443z

[28] Saksela M, Lapatto R, Raivio KO. Xanthine oxidoreductase gene expression and enzyme activity in developing human tissues. Biology of the Neonate. 1998;**74**:274-280. DOI: 10.1159/000014034

[29] Stirpe F, Della Corte E. The regulation of rat liver xanthine oxidase. Conversion in vitro of the enzyme activity from dehydrogenase (type D) to oxidase (type O). Journal of Biological Chemistry. 1969;**244**:3855-3863

[30] Kuwabara Y, Nishino T, Okamoto K, Matsumura T, Eger BT, Pai EF, et al. Unique amino acids cluster for switching from the dehydrogenase to oxidase form of xanthine oxidoreductase. Proceedings of the National Academy of Sciences of the United States of America. 2003;**100**:8170-8175. DOI: 10.1073/pnas.1431485100

[31] Nishino T, Okamoto K, Kawaguchi Y, Hori H, Matsumura T, Eger BT, et al. Mechanism of the conversion of xanthine dehydrogenase to xanthine oxidase: Identification of the two cysteine disulfide bonds and crystal structure of a non-convertible rat liver xanthine dehydrogenase mutant. The Journal of Biological Chemistry. 2005;**280**:24888-24894. DOI: 10.1074/jbc.M501830200

[32] Godber BLJ, Doel JJ, Sapkota GP, Blake DR, Stevens CR, Eisenthal R, et al. Reduction of nitrite to nitric oxide catalyzed by xanthine oxidoreductase. The Journal of Biological Chemistry. 2000;**275**:7757-7763. DOI: 10.1074/jbc.275.11.7757

[33] Maia LB, Moura JJG. Nitrite reduction by xanthine oxidase family enzymes: A new class of nitrite reductases. Journal of Biological Inorganic Chemistry. 2011;**16**:443-460. DOI: 10.1007/s00775-010-0741-z

[34] Li H, Cui H, Kundu TK, Alzawahra W, Zweier JL. Nitric oxide production from nitrite occurs primarily in tissues not in the blood: Critical role of xanthine oxidase and aldehyde oxidase. The Journal of Biological Chemistry. 2008;**283**:17855-17863. DOI: 10.1074/jbc.M801785200

[35] Li H, Samouilov A, Liu X, Zweier JL. Characterization of the magnitude and kinetics of xanthine oxidase-catalyzed nitrite reduction. The Journal of Biological Chemistry. 2001;**276**:24482-24489. DOI: 10.1074/jbc.m011648200

[36] Zweier JL, Samouilov A, Kuppusamy P. Non-enzymatic nitric oxide synthesis in biological systems. Biochimica Biophysica Acta – Bioenergy. 1999;**1411**:250-262. DOI: 10.1016/S0005-2728(99)00018-3

[37] Kovac S, Dinkova-Kostova A, Abramov A. The role of reactive oxygen species in epilepsy. React Oxyg Species. 2016;**1**:38-52. DOI: 10.20455/ros.2016.807

[38] Fang FC. Antimicrobial reactive oxygen and nitrogen species: Concepts and controversies. Nature Reviews. Microbiology. 2004;**2**:820-832. DOI: 10.1038/nrmicro1004

[39] Kelley EE, Hock T, Khoo NKH, Richardson GR, Johnson KK, Powell PC, et al. Moderate hypoxia induces xanthine oxidoreductase activity in arterial endothelial cells. Free Radical Biology & Medicine. 2006;**40**:952-959. DOI: 10.1016/j.freeradbiomed.2005.11.008

[40] Zhou L, Stanley WC, Saidel GM, Yu X, Cabrera ME. Regulation of lactate

production at the onset of ischaemia is independent of mitochondrial NADH/NAD+: Insights from in silico studies. The Journal of Physiology. 2005;**569**:925-937. DOI: 10.1113/jphysiol.2005.093146

[41] Kelley EE. A new paradigm for XOR-catalyzed reactive species generation in the endothelium. Pharmacol Reports. 2015;**67**:669-674. DOI: 10.1016/j.pharep.2015.05.004

[42] Robert AM, Robert L. Xanthine oxido-reductase, free radicals and cardiovascular disease. A critical review. Pathological and Oncological Research. 2014;**20**:1-10. DOI: 10.1007/s12253-013-9698-x

[43] Barsotti C, Ipata PL. Metabolic regulation of ATP breakdown and of adenosine production in rat brain extracts. The International Journal of Biochemistry & Cell Biology. 2004;**36**:2214-2225. DOI: 10.1016/j.biocel.2004.04.015

[44] Kelley EE, Khoo NKH, Hundley NJ, Malik UZ, Freeman B, a., Tarpey MM. Hydrogen peroxide is the major oxidant product of xanthine oxidase. Free Radical Biology & Medicine. 2010;**48**:493-498. DOI: 10.1016/j.freeradbiomed.2009.11.012

[45] Fatokun AA, Stone TW, Smith RA. Hydrogen peroxide mediates damage by xanthine and xanthine oxidase in cerebellar granule neuronal cultures. Neuroscience Letters. 2007;**416**:34-38. DOI: 10.1016/j.neulet.2007.01.078

[46] Walling C. Fenton's reagent revisited. Accounts of Chemical Research. 1975;**8**:125-131. DOI: 10.1021/ar50088a003

[47] Hausladen A, Fridovich I. Superoxide and peroxynitrite inactivate aconitases, but nitric oxide does not. The Journal of Biological Chemistry. 1994;**269**:29405-29408. DOI: 10.1016/s0021-9258(18)43893-8

[48] Franco MC, Ye Y, Refakis CA, Feldman JL, Stokes AL, Basso M, et al. Nitration of Hsp90 induces cell death. Proceedings of the National Academy of Sciences of the United States of America. 2013;**110**:E1102-E1111. DOI: 10.1073/pnas.1215177110

[49] Szabo C, Ischiropoulos H, Radi R. Peroxynitrite: Biochemistry, pathophysiology and development of therapeutics. Nature Reviews. Drug Discovery. 2007;**6**:662-680. DOI: 10.1038/nrd2222

[50] Barrera G, Pizzimenti S, Dianzani MU. Lipid peroxidation: Control of cell proliferation, cell differentiation and cell death. Molecular Aspects of Medicine. 2008;**29**:1-8. DOI: 10.1016/j.mam.2007.09.012

[51] Esterbauer H, Schaur RJ, Zollner H. Chemistry and biochemistry of 4-hydroxynonenal, malonaldehyde and related aldehydes. Free Radical Biology & Medicine. 1991;**11**:91-128. DOI: 10.1016/0891-5849(91)90192-6

[52] Kovac S, Dinkova-Kostova AT, Abramov AY. The role of reactive oxygen species in epilepsy. React Oxyg Species. 2005;**15**:273-277. DOI: 10.1080/13537909308580698

[53] Abramov AY, Scorziello A, Duchen MR. Cellular/molecular three distinct mechanisms generate oxygen free radicals in neurons and contribute to cell death during anoxia and reoxygenation. The Journal of Neuroscience. 2007;**27**:1129-1138. DOI: 10.1523/JNEUROSCI.4468-06.2007

[54] Nishino T. The conversion of xanthine dehydrogenase to xanthine oxidase and the role of the enzyme in reperfusion injury. Journal

of Biochemistry. 1994;**116**:1-6. DOI: 10.1093/oxfordjournals.jbchem.a124480

[55] Murad F. Nitric oxide and cyclic GMP in cell signaling and drug development. Nitric Oxide. 2006;**355**:2003-2011. DOI: 10.1056/NEJMsa063904

[56] Marla SS, Lee J, Groves JT. Peroxynitrite rapidly permeates phospholipid membranes. Proceedings of the National Academy of Sciences of the United States of America. 1997;**94**:14243-14248. DOI: 10.1073/pnas.94.26.14243

[57] Macmillan-Crow LA, Crow JP, Kerby JD, Beckman JS, Thompson JA. Nitration and inactivation of manganese superoxide dismutase in chronic rejection of human renal allografts. Proceedings of the National Academy of Sciences of the United States of America. 1996;**93**:11853-11858. DOI: 10.1073/pnas.93.21.11853

[58] Ghosh SM, Kapil V, Fuentes-Calvo I, Bubb KJ, Pearl V, Milsom AB, et al. Enhanced vasodilator activity of nitrite in hypertension: Critical role for erythrocytic xanthine oxidoreductase and translational potential. Hypertension. 2013;**61**:1091-1102. DOI: 10.1161/HYPERTENSIONAHA.111.00933

[59] Salgo MG, Bermudez E, Squadrito GL, Pryor WA. DNA damage and oxidation of thiols Peroxynitrite causes in rat thymocytes. Archives of Biochemistry and Biophysics. 1995;**322**:500-505. DOI: 10.1006/abbi.1995.1493

[60] Quijano C, Alvarez B, Gatti RM, Augusto O, Radi R. Pathways of peroxynitrite oxidation of thiol groups. The Biochemical Journal. 1997;**322**:167-173. DOI: 10.1042/bj3220167

[61] Baker JF, Krishnan E, Chen L, Schumacher HR. Serum uric acid and cardiovascular disease: Recent developments, and where do they leave us? The American Journal of Medicine. 2005;**118**:816-826. DOI: 10.1016/j.amjmed.2005.03.043

[62] Scheepers LE, Wei F, Stolarz-Skrzypek K, Malyutina S, Tikhonoff V, Thijs L, et al. THU0519 xanthine oxidase gene variants and their association with blood pressure and incident hypertension: A population study. Annals of the Rheumatic Diseases. 2016;**75**:379-380. DOI: 10.1136/annrheumdis-2016-eular.2964

[63] Guarda NS, Bollick YS, De Carvalho JAM, Premaor MO, Comim FV, Moresco RN. High serum uric acid is associated with tubular damage and kidney inflammation in patients with type 2 diabetes. Disease Markers. 2019;**2019**:1-9. DOI: 10.1155/2019/6025804

[64] Annanmaki T, Muuronen A, Murros K. Low plasma uric acid level in Parkinson's disease. Movement Disorders. 2007;**22**:1133-1137. DOI: 10.1002/mds.21502

[65] Auinger P, Kieburtz K, McDermott MP. The relationship between uric acid levels and Huntington's disease progression. Movement Disorders. 2010;**25**:224-228. DOI: 10.1002/mds.22907

[66] Ames BN, Cathcart R, Schwiers E, Hochstein P. Uric acid provides an antioxidant defense in humans against oxidant- and radical-caused aging and cancer: A hypothesis. Proceedings of the National Academy of Sciences of the United States of America. 1981;**78**:6858-6862. DOI: 10.1073/pnas.78.11.6858

[67] Perez-Ruiz F, Dalbeth N, Schlesinger N. Febuxostat, a novel drug for the treatment of hyperuricemia of gout. Future Rheumatology. 2008;**3**:421. DOI: 10.2217/17460816.3.5.421

[68] Sawa T, Wu J, Akaike T, Maeda H. Tumor-targeting chemotherapy by a xanthine oxidase-polymer conjugate that generates oxygen-free radicals in tumor tissue. Cancer Research. 2000;**60**:666-671

[69] Yu MA, Sánchez-Lozada LG, Johnson RJ, Kang DH. Oxidative stress with an activation of the renin-angiotensin system in human vascular endothelial cells as a novel mechanism of uric acid-induced endothelial dysfunction. Journal of Hypertension. 2010;**28**:1234-1242. DOI: 10.1097/HJH.0b013e328337da1d

[70] Klauser AS, Halpern EJ, Strobl S, Gruber J, Feuchtner G, Bellmann-Weiler R, et al. Dual-energy computed tomography detection of cardiovascular monosodium urate deposits in patients with gout. JAMA Cardiology. 2019;**4**:1019-1028. DOI: 10.1001/jamacardio.2019.3201

[71] Yoshida S, Kurajoh M, Fukumoto S, Murase T, Nakamura T, Yoshida H, et al. Association of plasma xanthine oxidoreductase activity with blood pressure affected by oxidative stress level: MedCity21 health examination registry. Scientific Reports. 2020;**10**:4437. DOI: 10.1038/s41598-020-61463-8

[72] Mori K, Furuhashi M, Tanaka M, Numata K, Hisasue T, Hanawa N, et al. U-shaped relationship between serum uric acid level and decline in renal function during a 10-year period in female subjects: BOREAS-CKD2. Hypertension Research. 2021;**44**:107-116. DOI: 10.1038/s41440-020-0532-z

[73] Yang J, Kamide K, Kokubo Y, Takiuchi S, Horio T, Matayoshi T, et al. Associations of hypertension and its complications with variations in the xanthine dehydrogenase gene. Hypertension Research. 2008;**31**:931-940. DOI: 10.1291/hypres.31.931

[74] Battelli MG, Polito L, Bortolotti M, Bolognesi A. Xanthine oxidoreductase in cancer: More than a differentiation marker. Cancer Medicine. 2016;**5**:546-557. DOI: 10.1002/cam4.601

[75] Battelli MG, Polito L, Bortolotti M, Bolognesi A. Xanthine oxidoreductase-derived reactive species: Physiological and pathological effects. Oxidative Medicine and Cellular Longevity. 2016;**2016**:1-8. DOI: 10.1155/2016/3527579

[76] Buetler TM, Krauskopf A, Ruegg UT. Role of superoxide as a signaling molecule. News in Physiological Sciences. 2004;**19**:120-123. DOI: 10.1152/nips.01514.2003

[77] Krylatov AV, Maslov LN, Voronkov NS, Boshchenko AA, Popov SV, Gomez L, et al. Reactive oxygen species as intracellular signaling molecules in the cardiovascular system. Current Cardiology Reviews. 2018;**14**:290-300. DOI: 10.2174/1573403x14666180702152436

[78] Igaki Y, Tanno M, Sato T, Kouzu H, Ogawa T, Osanami A, et al. Xanthine oxidoreductase-mediated injury is amplified by upregulated AMP deaminase in type 2 diabetic rat hearts under the condition of pressure overload. Journal of Molecular and Cellular Cardiology. 2021;**154**:21-31. DOI: 10.1016/j.yjmcc.2021.01.002

[79] Battelli MG, Bolognesi A, Polito L. Pathophysiology of circulating xanthine oxidoreductase: New emerging roles

for a multi-tasking enzyme. Biochimica Biophysica Acta – Molecular Basis Diseases. 2014;**1842**:1502-1517. DOI: 10.1016/j.bbadis.2014.05.022

[80] Linder N, Lundin J, Isola J, Lundin M, Raivio KO, Joensuu H. Down-regulated xanthine oxidoreductase is a feature of aggressive breast cancer. Clinical Cancer Research. 2005;**11**:4372-4381. DOI: 10.1158/1078-0432.CCR-04-2280

[81] Fini MA, Orchard-Webb D, Kosmider B, Amon JD, Kelland R, Shibao G, et al. Migratory activity of human breast cancer cells is modulated by differential expression of xanthine oxidoreductase. Journal of Cellular Biochemistry. 2008;**105**:1008-1026. DOI: 10.1002/jcb.21901

[82] Linder N, Haglund C, Lundin M, Nordling S, Ristimaki A, Kokkola A, et al. Decreased xanthine oxidoreductase is a predictor of poor prognosis in early-stage gastric cancer. Journal of Clinical Pathology. 2006;**59**:965-971. DOI: 10.1136/jcp.2005.032524

[83] Linder N, Buetzow R, Lassus H, Lundin M, Lundin J. Decreased xanthine oxidoreductase (XOR) is associated with a worse prognosis in patients with serous ovarian carcinoma. Gynecologic Oncology. 2012;**124**:311-318. DOI: 10.1016/j.ygyno.2011.10.026

[84] Linder N, Martelin E, Lundin M, Louhimo J, Nordling S, Haglund C, et al. Xanthine oxidoreductase – clinical significance in colorectal cancer and in vitro expression of the protein in human colon cancer cells. European Journal of Cancer. 2009;**45**:648-655. DOI: 10.1016/j.ejca.2008.10.036

[85] Xu H, Li C, Mozziconacci O, Zhu R, Xu Y, Tang Y, et al. Xanthine oxidase-mediated oxidative stress promotes cancer cell-specific apoptosis. Free Radical Biology & Medicine. 2019;**139**:70-79. DOI: 10.1016/J.FREERADBIOMED.2019.05.019

[86] Choi HK, Mount DB, Reginato AM. Pathogenesis of gout. Annals of Internal Medicine. 2005;**143**:499-516. DOI: 10.7326/0003-4819-143-7-200510040-00009

[87] Okamoto K, Eger BT, Nishino T, Kondo S, Pai EF, Nishino T. An extremely potent inhibitor of xanthine oxidoreductase. The Journal of Biological Chemistry. 2003;**278**:1848-1855. DOI: 10.1074/jbc.M208307200

[88] Matsumoto K, Okamoto K, Ashizawa N, Nishino T. FYX-051: A novel and potent hybrid-type inhibitor of xanthine oxidoreductase. The Journal of Pharmacology and Experimental Therapeutics. 2011;**336**:95-103. DOI: 10.1124/jpet.110.174540

[89] Diaz-Torné C, Perez-Herrero N, Perez-Ruiz F. New medications in development for the treatment of hyperuricemia of gout. Current Opinion in Rheumatology. 2015;**27**:164-169. DOI: 10.1097/BOR.0000000000000146

[90] Corran HS, Dewan JG, Gordon AH, Green DE. Xanthine oxidase and milk flavoprotein: With an Addendum by J. St L. Philpot. Biochemical Journal. 1939;**33**:1694-1708

Chapter 5

Methylmercury Promotes Oxidative Stress and Activation of Matrix Metalloproteinases: Cardiovascular Implications

Keuri Eleutério Rodrigues, Stefanne de Cássia Pereira da Silva and Alejandro Ferraz do Prado

Abstract

Preclinical and clinical studies worldwide have shown an association between methylmercury (MeHg) poisoning and the risk of developing cardiovascular diseases such as arrhythmias, arterial hypertension, atherosclerosis and myocardial infarction. One of the hypotheses raised for MeHg-induced toxicity is associated with redox imbalance, which promotes oxidative stress by increasing reactive oxygen species (ROS) and reducing the activity of antioxidant enzymes such as catalase (CAT), superoxide dismutase (SOD) and glutathione peroxidase (GPx). In addition, oxidative stress and organomercurial compounds are capable of activating MMPs. MMP-2 and MMP-9 participate in pathophysiological processes associated with cardiovascular remodeling. A positive correlation between mercury exposure and increased plasma activity of MMP-2 and circulating MMP-9 has been demonstrated, suggesting a possible mechanism that could increase susceptibility to cardiovascular diseases.

Keywords: mercury, MMP-2, remodeling, redox state, cardiovascular dysfunction

1. Introduction

Mercury (Hg), an environmental pollutant from natural and anthropogenic sources, is converted into methylmercury (MeHg), a more toxic organic form capable of bioaccumulating in food webs [1, 2]. Humans are exposed to MeHg through the consumption of contaminated fish, especially predatory species, which through trophic magnification, accumulate higher amounts of the metal [3, 4]. The effects of human poisoning by MeHg took worldwide repercussions after the contamination of the Minamata Basin in Japan (1956). Years later, people who consumed fish from this region developed a "Minamata disease" syndrome, mainly affecting the nervous system [5].

Human exposure to MeHg can pose various health risks. The spectrum of adverse effects associated with MeHg poisoning will depend on the exposure time and magnitude of the dose [6, 7]. Life span also influences the damage caused by MeHg, and prenatal exposure can potentially cause irreversible damage to the developing central nervous system [8, 9]. The exposure that occurs during childhood and adulthood can also cause damage to the central nervous system. However, signs of toxicity appear months after the onset of exposure [10, 11].

In addition to affecting the Central Nervous System, MeHg compromises other systems [12–14]. For example, the repercussions of MeHg on the cardiovascular system have been investigated in recent decades [15–17]. A study of 1833 Finnish men aged 42–60 suggested a correlation between mercury accumulation in the body from high consumption of MeHg-contaminated non-fatty fish and the increased risk of acute myocardial infarction, coronary heart disease and other cardiovascular diseases. In this study, the researchers theorized that the effects of MeHg on the cardiovascular system could be associated with lipid peroxidation triggered by mercury [18].

One of the mechanisms involved in MeHg-induced cytotoxicity is the dysregulation of the redox state by increasing reactive species and suppressing the antioxidant system, triggering oxidative stress [19]. Furthermore, the membranes of cells and organelles are affected by the direct action of reactive species, which promote lipid peroxidation and generate changes in structure and permeability, which culminate in cell death [7, 20, 21]. In addition, MeHg promotes protein denaturation, enzyme inactivation, DNA damage and triggering epigenetic changes [22, 23].

Reactive species derived from oxygen and nitrogen have a crucial effect on the pathophysiology of cardiovascular diseases [24, 25]. These molecules activate matrix metalloproteinases (MMPs) [26–28]. The activation of MMPs can also occur directly by organomercurial compounds, which disrupt the interaction of a propeptide domain cysteine residue with catalytic zinc within the enzyme's active site [29, 30]. MMPs are a family of zinc-dependent endopeptidases that control the synthesis and degradation of extracellular matrix (ECM) components and can also act directly on intracellular substrates on a small-time scale [31, 32].

The group of gelatinases (MMP-2 and MMP-9) is extensively investigated in cardiovascular diseases. The increase in proteolytic activity is closely related to the remodeling of the heart and vessels [33–35]. It has also been suggested the existence of a correlation between increased MMP-9 and MMP-2 activity and plasma mercury concentrations as a possible mechanism that could increase susceptibility to cardiovascular diseases, showing a possible causal relationship between mercurial exposure and increased susceptibility to cardiovascular diseases [36].

2. Forms of mercury

The chemical element Mercury (Hg) is a heavy metal whose symbol Hg derives from the Latin hydrargyrum, meaning liquid silver. It belongs to the transition metals in the sixth period and family II B of the periodic table. Under average temperature and pressure conditions, it presents a liquid physical state with silver coloration. Its fluidity at room temperature, and high thermal and electrical conductivity, make it an excellent conductive material. Furthermore, its uniform volumetric expansion over a wide temperature range and high density makes it an ideal element for manufacturing instruments for physical measurements, such as thermometers, barometers and electrical systems.

Among the natural sources emitting mercury in the environment are: volcanic eruptions, degassing of the Earth's crust and mercury (HgS) mines [37–39].

The anthropogenic sources of mercury emission are multiple, emphasizing the chemical industry, with the burning of fossil fuels, the production of electronics, such as batteries, and the amalgamation of mercury used in dentistry and gold mining [40–42].

Mercury is widely distributed in nature, forming several compounds that are grouped into three groups: elemental mercury (Hg^0), the inorganic species: cinnabar ore, mercury oxide, mercurous ion and mercuric ion (HeS, HgO, $Hg2^{2+}$, Hg^{2+}) respectively and the organic species, most common dimethylmercury ($(CH_3)_2Hg$) and methylmercury (CH_3Hg^+). When mercury combines with carbon, it forms so-called organic or organomercurial compounds. The conversion of mercury to methylmercury (MeHg) occurs mainly in aquatic systems and can also be converted in soil and sediments in a smaller proportion [12, 43].

3. The mercury cycle in the environment and the methylation process

Understanding the dynamics of mercury in the environment is very important for understanding its actual impact on living things. The mercury cycle is characterized by the various routes that this compound can follow in the ambient carried by human or natural activity to the biosphere, transiting in different forms and states of oxidation, its distribution in the separate compartments that make up the environment: water, air and soil, produces extremely varied distribution patterns [44, 45].

In the atmosphere, the elemental mercury (Hg^0), in the form of vapor, oxidizes, giving rise to the intermediate state of bivalent ionic mercury (Hg^{2+}), which then complexes the molecule of ozone or chlorine, giving rise to substances (Hg^O and $HgCl_2$) less volatile and more soluble in water, which are easily solubilized in the water vapor of the clouds, which are subsequently deposited on the vegetation, soil and water. The Hg^{2+} also undergoes precipitation with rain, the most common form found in soil and surface water, and may undergo methylation or volatilize and return to the atmosphere [44, 45].

In the soil, the mercury deposited in the bivalent form can be complex to humic acids, clay mineral particles and mainly organic matter, conferring stability and high capacity to keep in the environment for long, even after eliminating the generating source. Furthermore, the mercury in the soil can be methylated or undergo sorption, converting into volatile forms that will return to the atmosphere or even dissolve in the aquatic environment, undergoing methylated and bioaccumulation in the food chain [44–46].

Inorganic mercury can be converted into elemental mercury in the aquatic ecosystem by microorganisms, humic and fulvic acids under specific physicochemical conditions. The major transformation undergone by inorganic mercury in the aquatic environment from the environmental point of view is its conversion into organomercurial compounds, among which methylmercury, the most toxic and bioavailable form, stands out. Surfactant bacteria interact with methylcobalamin, known as vitamin B12, transferring the methyl group to mercury, generating methylmercury [47, 48]. The synthesis rate of MeHg considers the composition of the species of bacteria, temperature, organic carbon, sulfur and dissolved oxygen for transfer of groups and acidity. Elemental mercury is little reactive in a natural aquatic environment, presenting a low possibility of oxidation, with practically zero contribution to the formation of MeHg [49, 50].

Methylmercury is stable in aquatic environments, forming several soluble complexes, thus favoring its dispersion. Methylmercury dissolved in water can be

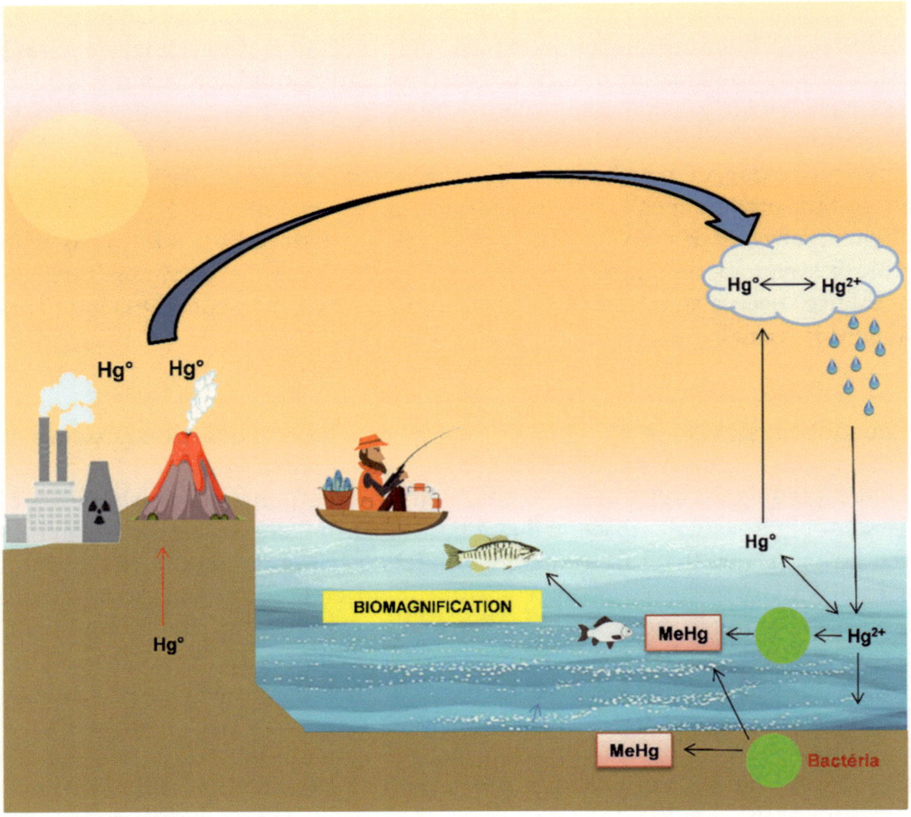

Figure 1.
Mercury cycle: Mercury can follow several routes in the different compartments that make up the environment, presenting various forms such as elemental mercury (Hgo), mercuric ion (Hg2+) and methylmercury (CH3Hg+), being released into the biosphere both by anthropogenic activity, burning of fossil fuel, and natural sources such as volcanic eruptions. In the atmosphere, Hgo, in the form of steam, can be converted into Hg2+ that is complex to other molecules, giving rise to substances that are associated with cloud water, precipitating with rain, depositing in water and soil, and may return to the atmosphere or undergo methylation by bacteria, and may be incorporated by plankton entering the food chain, suffers trophic magnification and bioaccumulation reaching organisms that are at the top of the food chain.

incorporated by plankton entering the food chain. Trophic magnification and bioaccumulation result in very high levels of MeHg in predatory fish, organisms at the top of the aquatic food chain, which are a significant source of exposure of wild animals and humans to MeHg [51–53] (**Figure 1**).

4. Toxic effects of methylmercury on human health

Throughout history, mercury has been used for various purposes, presenting a wide distribution in nature, resulting in intoxication in animal and human populations. However, the main form of human exposure to MeHg occurs from the diet, mainly from consuming contaminated fish [54–58].

Due to its liposolubility, MeHg ingested in food and rapidly absorbed by the gastrointestinal tract easily crosses the membranes of cells, the placenta and the blood-brain barrier [59, 60]. In addition, the systems in development are affected more severely due to the period of differentiation and cell migration, generating irreversible damage compared to adult organisms [61–63]. Therefore, populations whose diet is associated with the consumption of fish and marine mammals are at immediate risk of MeHg poisoning [54–58, 64].

MeHg is absorbed by the gastrointestinal tract and enters the bloodstream and bioaccumulates in different organs such as the liver, kidney, lung, brain and heart [65]. Because it is a metal with an electrophilic nature, MeHg reacts with nucleophiles such as selenol (SeH) and sulfhydryl (-SH) groups, forming stable complexes. In the intracellular environment, MeHg can form these complexes with proteins, as well as with GSH and the amino acid cysteine (Cys), which have these groups in their structures [66–68].

Conjugated MeHg circulates more easily in the blood and has greater permeability and distribution, as it now has a molecular structure similar to endogenous molecules [69]. Some studies have suggested that the complex formed by methylmercury-L-cysteine (MeHg-Cys) has a molecular structure identical to the amino acid methionine (Met), which crosses biomembranes through the L-type amino acid transporter (LAT1, LAT2 and LAT3), with However, the physiological function of this transporter still needs to be clarified, as well as its subcellular location, for a better understanding of MeHg toxicodynamics [70–72].

MeHg is rapidly taken into the intracellular medium, retained in subcellular regions, and slowly converted into inorganic mercury [73]. Available treatments for heavy metal poisoning use chelating agents such as 2,3-dimercaptopropanol (BAL), meso-2,3-dimercaptosuccinic (DMSA) and 2,3-dimercaptopropane-1-sulfonate (DMPS). However, these molecules are ineffective in MeHg detoxification, as they have a low therapeutic window, unable to remove the metal at the intracellular level. They can also cause the redistribution of the same in the body, often causing hepatotoxicity and nephrotoxicity, making it difficult to implement an appropriate treatment therapy in cases of MeHg intoxication [74–76].

The exact mechanism of cytotoxicity induced by MeHg is still not fully understood. However, many studies suggest that changes in the antioxidant system, calcium homeostasis and the glutamatergic system are involved in the toxic effects at the cellular level generated by MeHg [66].

In the Japanese city of Minamata in 1956, people who consumed fish and shellfish contaminated with MeHg, dumped by a chemical factory, developed a syndrome known as Minamata disease. People developed sensory disturbances, ataxia, dysarthria, visual field constriction, auditory changes and tremors. In addition, pregnant women in the region, who feed on species of this contaminated ecosystem, had children with extensive brain lesions [5, 11].

In Iraq, around 1971/1972, people were poisoned with MeHg by contamination of seeds with organomercurial compounds used as a fungicide. In addition, the consumption of homemade products prepared from these seeds caused the poisoning of many people, who presented as the most common symptom of paresthesia. In the most severe cases, the individuals exhibited ataxia, changes in vision, speech and hearing, followed by blindness, deafness and death [77, 78].

The two episodes of human poisoning by MeHg from consuming contaminated food served as bases to determine the effects of methylmercury. It was possible to establish that high doses of this metal in the body affect the nervous system and even lead to death. Furthermore, methylmercury could cross the transplacental barrier and cause irreversible damage to the fetus [11, 77, 79, 80].

However, human exposure to methylmercury generally occurs chronically and at low doses, and the spectrum of adverse effects and severity of damage largely depends on the magnitude of the exposure dose [81]. Children exposed to high amounts of methylmercury while still in the womb, as in the cases of Minamata and Iraq, were born with severe disabilities such as mental retardation, microcephaly, seizures, and blindness [82–84]. Exposure in utero to low doses of MeHg can generate effects that go unnoticed at birth. However, during development, the child presents neurocognitive deficits, such as difficulties in crawling, sitting, walking and learning [85].

Recently, epidemiological studies have been conducted to assess the human health risks of chronic exposure to low doses of MeHg. Two extensive studies that addressed the effects of MeHg on child development: the Seychelles Child Development Study (SCDS) and the Ilas Faroe Study, obtained divergent results. In the first SCDS, no adverse effects were found, while in the second Faroe Islands, a range of adverse effects was described, among them neuropsychological and neurophysiological impairment [86–88].

The effects caused by different doses of MeHg on the nervous system of adults and children have been extensively investigated over the last decades [89]. However, there is evidence that exposure to MeHg promotes changes in the cardiovascular system generating dysregulation of blood pressure, changes in heart rate and increased risk of developing coronary heart disease, atherosclerosis and acute myocardial infarction in adults [90–92].

5. Cardiovascular effects of MeHg and the role of oxidative stress, MMP-2 and MMP-9

MeHg is the major organic form of mercury involved in poisoning events in humans. It has relevant cardiovascular effects flagged in 2000 by the National Research Council on the Toxicological Effects of MeHg through a report that reinforced the need for more research. Clinical studies have observed associations between mercury levels and increased risk for myocardial infarction in Finland [18, 93, 94], Israel and eight European countries [95]. In addition, associations between mercury and arterial hypertension, vascular hypertrophy and arrhythmias were found in population studies in Wisconsin, the Brazilian Amazon region and among members of Faroese Whaling men [90–92].

Studies in cell and animal cultures have been performed to understand the cardiovascular toxicodynamic mechanisms of MeHg and demonstrated time-dependent effects. In acute exposure in rats, increased vascular relaxation was observed, with no changes in blood pressure [96]. On the other hand, in times of greater exposure, weight loss, mortality, hypertension, hypercholesterolemia, vascular hypertrophy, inflammation and atherosclerotic lesions and cardiac arrhythmias are observed [97–101]. Vascular findings have been associated with oxidative stress, inflammation and endothelial dysfunction [100, 101]. In fact, studies in endothelial cell culture have demonstrated that MeHg decreases cell viability [102–104] by the production of ROS and activation of phospholipases. These effects are prevented by the use of antioxidants [102, 104]. Subsequently, it was demonstrated that ROS production depends on the activation of NADPH oxidase and the decrease in antioxidant defense [105].

MeHg can bind directly to the tripeptide GSH by a sulfhydryl bond with the amino acid cysteine forming a complex (MeHg-GSH) that can be exported from the cell leading to depletion of its intracellular levels. However, this is not the only mechanism involved in reducing intracellular GSH levels. The greater interaction affinity of MeHg with selenocysteine residues present in GPx and the inhibition of its expression and activity alter the GSH-GSSG redox cycle, thus decreasing the GSH conversion rate [106, 107]. MeHg also can increase superoxide anion and hydrogen peroxide generation via NADPH oxidase and mitochondria, leading to intracellular GSH depletion. There is a dichotomous relationship between MeHg and GSH because this interaction favors MeHg excretion.

On the other hand, this interaction triggers cytotoxicity mechanisms. In addition, MeHg also decreases the activity of other antioxidant enzymes, such as SOD and catalase, contributing to increased ROS bioavailability [66, 97, 105, 108, 109]. In this way, MeHg induces oxidative stress that directly contributes to endothelial dysfunction.

Endothelial dysfunction is a key event in cardiovascular diseases, including hypertension, atherosclerosis and cardiomyopathies [110–112]. The leading cause and consequence of the loss of endothelial functionality is the decrease in nitric oxide (NO) production. NO can be produced by endothelial nitric oxide synthetase (eNOS), neuronal oxide synthetase (nNOS) and induced oxide synthetase (iNOS) that perform the conversion of the amino acid L-arginine and molecular oxygen into nitric oxide and L-citrulline. The reaction requires the presence of co-factors such as NADPH (nicotinamide adenine dinucleotide phosphate), BH4 (tetrahydrobiopterin), flavin adenine dinucleotide (FAD) and flavin mononucleotide (FMN) [113]. The main factor involved in the decrease in the bioavailability of NO is oxidative stress, resulting from the overproduction and inactivation of ROS such as superoxide (O_2^-) leads to rapid chemical reaction with NO forming peroxynitrite ($ONOO^-$), implying lower bioavailability of NO contributing to cardiovascular events [114]. Low MeHg concentrations induce hypertension in rats, which has been associated with oxidative stress and lower bioavailability of NO [97]. In addition to oxidative stress, MeHg generates S-mercuration, a post-translational modification in which mercury reacts with nucleophiles, such as the thiol group of proteins, leading to structural and functional alteration of molecules. MeHg has been described to cause S-mercuration of superoxide dismutase (SOD) and nNOS, thereby decreasing ROS inactivation and NO production [115, 116].

Organomercurial compounds such as MeHg can alter cardiovascular homeostasis also by modulating the activity of MMPs [29, 30]. MMPs control the content of the extracellular matrix and comprise a group of 28 proteins, 24 of which are found in humans. MMPs are classified according to their structure and primary degradation substrate. For example, MMP-2 and MMP-9 are present in the gelatinase group, the main MMPs studied in the cardiovascular system, mainly due to their relevant role and ease of evaluation with laboratory techniques. MMP-2 and MMP-9, similar to the other MMPs, have a signal peptide, a propeptide, a catalytic site containing a zinc atom and a hemopexin domain. However, they differ from the rest of the other enzymes by having three domains of fibronectin at the catalytic site, responsible for the high affinity for denatured collagen (gelatin) [117] (**Figure 2**).

MMP-2 and MMP-9 expressed are secreted into the extracellular medium as the zymogen, called Pro-MMP-2 (72 kDa) and Pro-MMP-9 (92 kDa). Enzymatic inactivity is maintained by a sulfhydryl bond between a cysteine in the propeptide and zinc at the catalytic site [117]. The activation process requires breaking the sulfhydryl bond to expose the catalytic site, known as the " cysteine switch" [118]. Briefly, the activation of gelatinases can occur by cleavage of the propeptide by other enzymes, leading to a decrease in the molecular weight of MMP-2 (64 kDa) and MMP-9 (83 kDa) or by non-proteolytic activation that can occur by disruption of the sulfhydryl bond by reactive species and mercurial compounds [26–30, 117]. The non-proteolytic activation maintains the molecular weight of the enzyme. However, the conformational change of the structure subsequently allows the propeptide's auto-cleavage, resulting in a decrease in molecular weight (**Figure 2**). The enzymatic

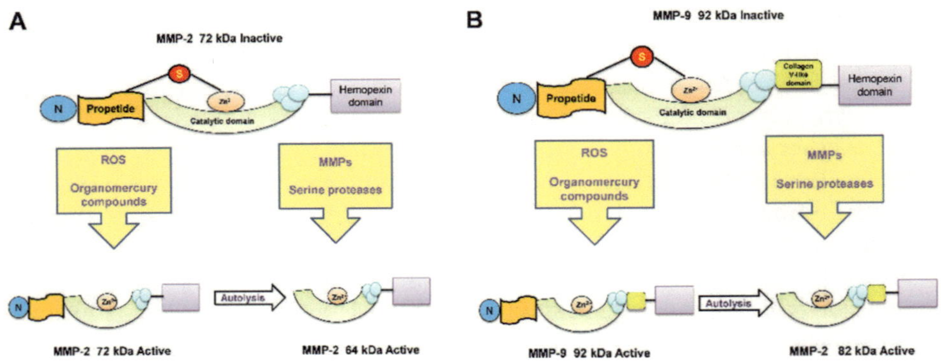

Figure 2.
Proteolytic and non-proteolytic activation of MMP-2 and MMP-9. (A) the 72 kDa MMP-2 (pro-MMP-2) has in its structure a signal peptide (N), a propeptide, a catalytic domain containing zinc and three fibronectin repeats and a hemopexin domain. The sulfhydryl bond between a cysteine residue in the propeptide domain with zinc ensures catalytic inactivity. The active 72 kDa MMP-2 can undergo non-proteolytic activation by ROS and organomercurial compounds by breaking the sulfhydryl bond, maintaining the structure and molecular weight of 72 kDa. The isoform of active 64 kDa MMP-2 lacks the propeptide and can be formed by the proteolytic action of MMPs or serine proteases and by autolysis. (B) the 92 kDa MMP-9 (pro-MMP-9) has in its structure a signal peptide (N), a propeptide, a catalytic domain containing zinc and three fibronectin repeats, a collagen-binding domain type V and hemopexin domain. The proteolytic and non-proteolytic activation processes are similar to those previously described.

activity of MMP-2 and MMP-9 is modulated by tissue inhibitors of MMPs (TIMPs) and alpha2-macroglobulin. The four TIMPs described are not selective regarding the inhibition of MMPs, but MMP-2 and MMP-9 are inhibited mainly by TIMP-2 and TIMP-1, respectively [117].

MeHg can directly activate MMP-2 and MMP-9, increasing tissue proteolytic activity [29, 30]. In fact, increased activity of MMP-2 and MMP-9 was observed in population samples from the Amazon region of Brazil, which showed high plasma levels of mercury due to a diet rich in contaminated fish. The authors demonstrated a negative correlation between mercury levels and TIMP-1 and TIMP-2. The positive correlation between mercury levels and the ratio MMP-9/TIMP-1 and MMP-2/TIMP-2 leads to increased enzyme gelatinolytic activity [36]. Later studies demonstrated that polymorphisms in the MMP-2 and MMP-9 genes are related to changes in gelatinase activity and MMP-2/9/TIMPs ratio in mercurial intoxication [119, 120]. It has also been shown that MeHg can epigenetically alter MMP-9 by inducing changes in cell junctions [22]. Together, these studies demonstrated biological mechanisms that may be related to the cardiovascular toxicity of MeHg (**Figure 3**).

The imbalance between the activity of MMPs and TIMPs is one of the main determinants of matrix composition and is related to increased cardiovascular risk [121, 122]. Studies have shown that increased MMP-2 activity in the heart leads to decreased cardiac systolic function due to the cleavage of sarcomere proteins, including troponin I, alpha-actinin, titin and light chain myosin [123–126]. In these studies, intracellular activation of MMP-2 by $ONOO^-$ was demonstrated to result in intracellular lysis of sarcomere proteins, resulting in severe cardiac damage. MMP-9 modulates inflammatory response by activating cytokines and chemokines, including TNFα, IL-1β, TGFβ and CXC-modified ligands [127–129]. In addition, MMP-2 and MMP-9 also activate profibrotic pathways in cardiovascular tissue [130–133]. It is worth

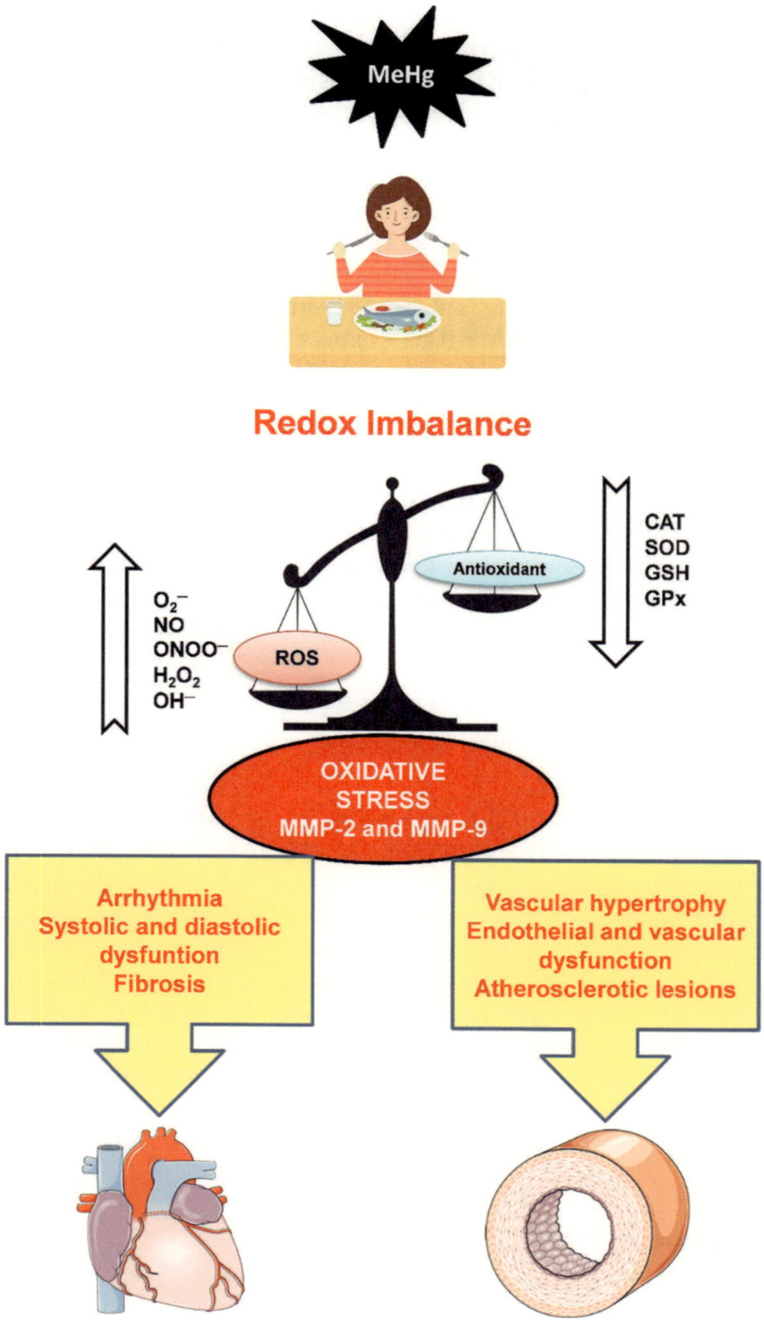

Figure 3.
Cardiovascular adverse events associated with MeHg poisoning: Humans become contaminated with MeHg from their diet, mainly by consuming contaminated fish. MeHg is ingested in food and rapidly absorbed by the gastrointestinal tract, crossing cell membranes, promoting cytotoxicity by altering the redox balance, generating an increase in reactive species and suppressing the antioxidant system, leading to oxidative stress. Oxidative stress and MeHg can activate MMP-2 and MMP-9. The higher proteolytic activities promote the remodeling and dysfunction of the heart and vessels, triggering adverse cardiovascular events.

mentioning that mercury is also stored in the heart [134–136], which can lead to direct activation of MMPs and disruption of the thiol-zinc bond, which can generate adverse events similar to those observed in the model of ischemia and reperfusion of the heart.

Although we have pointed out evidence of the toxic cardiovascular effects of MeHg, some population studies have not found a correlation between mercury levels and cardiovascular changes [10, 91, 92, 137, 138]. Furthermore, despite the signaling that MeHg induces oxidative stress and that both directly activate MMPs, few studies demonstrate the participation of MMPs in cardiovascular toxicodynamic events associated with MeHg. However, this triple association resulted in functional alterations in other body systems. Decreased renal function by cytoskeleton disruption was associated with increased MMP-9 gene expression via oxidative stress induced by MeHg [22]. In another study, activation of MMP-9 and MMP-13 via MeHg-induced redox alteration led to altered embryonic development in zebrafish [139]. Neurodevelopmental alterations caused by MeHg have also been associated with gene alterations of MMP-1 and antioxidant enzymes [140].

In summary, we show evidence that MeHg can modulate directly or indirectly via oxidative stress the expression and activity of MMPs. As previously mentioned, small alterations in MMP activity can lead to serious cardiovascular alterations. Thus, the interaction of MeHg, oxidative stress and MMPs can be a point to be looked at more thoroughly, and this could be a mechanism involved in the cardiotoxicity of MeHg.

6. Conclusions

The studies discussed in this chapter have shown that MeHg poisoning, both in low doses and in high doses, in animal models and humans, is capable of causing adverse events in the cardiovascular system. The cytotoxicity induced by MeHg causes depletion of the antioxidant system, both of molecules such as glutathione, as well as of antioxidant enzymes (GPx, SOD and CAT) and increase of reactive species (superoxide and NO). Furthermore, MeHg and oxidative stress can alter the expression and activity of MMP-2 and MMP-9, suggesting a possible mechanism of susceptibility to cardiovascular diseases. However, well-defined epidemiological studies are needed to establish a cause-and-effect relationship between dietary MeHg poisoning associated with altered redox status and a correlation between the increased activity of MMP-2 and MMP-9 with increased developmental richness with cardiovascular diseases.

Acknowledgements

This work had partial financial support from Conselho Nacional de Desenvolvimento científico e Tecnológico (CNPq 427591/2018-0), Coordenação de Aperfeiçoamento de Pessoal de Nível Superior—Brasil (CAPES) —Finance Code 001. The APC payment was supported by Pró-reitoria de Pesquisa e Pós-Graduação (PROPESP) from the Federal University of Pará (UFPA).

Conflict of interest

The authors declare no conflict of interest.

Author details

Keuri Eleutério Rodrigues[1], Stefanne de Cássia Pereira da Silva[1,2] and Alejandro Ferraz do Prado[1]*

1 Laboratory of Pharmacology and Toxicology of the Cardiovascular System, Institute of Biological Sciences, Federal University of Pará, Belém, Brazil

2 Center for Biological and Health Sciences, University of the State of Pará, Belém, Brazil

*Address all correspondence to: alejandrofp@ufpa.br

References

[1] Hilgendag IR et al. Mercury biomagnification in benthic, pelagic, and benthopelagic food webs in an Arctic marine ecosystem. Science of the Total Environment. 2022;**841**:156424

[2] Zhang F et al. Terrestrial mercury and methylmercury bioaccumulation and trophic transfer in subtropical urban forest food webs. Chemosphere. 2022;**299**:134424

[3] Carneiro MF, Grotto D, Barbosa F Jr. Inorganic and methylmercury levels in plasma are differentially associated with age, gender, and oxidative stress markers in a population exposed to mercury through fish consumption. Journal of Toxicology and Environmental Health. Part A. 2014;**77**(1-3):69-79

[4] Salazar-Camacho C et al. A human health risk assessment of methylmercury, arsenic and metals in a tropical river basin impacted by gold mining in the Colombian Pacific region. Environmental Research. 2022;**212**(Pt B):113120

[5] Harada M. Minamata disease: Methylmercury poisoning in Japan caused by environmental pollution. Critical Reviews in Toxicology. 1995;**25**(1):1-24

[6] Karagas MR et al. Evidence on the human health effects of low-level methylmercury exposure. Environmental Health Perspectives. 2012;**120**(6):799-806

[7] Yang T et al. Oxidative stress accelerates synaptic glutamate dyshomeostasis and NMDARs disorder during methylmercury-induced neuronal apoptosis in rat cerebral cortex. Environmental Toxicology. 2020;**35**(6):683-696

[8] Llop S et al. Prenatal exposure to mercury and neuropsychological development in young children: The role of fish consumption. International Journal of Epidemiology. 2017;**46**(3):827-838

[9] van de Bor M. Fetal toxicology. Handbook of Clinical Neurology. 2019;**162**:31-55

[10] Grandjean P et al. Cardiac autonomic activity in methylmercury neurotoxicity: 14-year follow-up of a Faroese birth cohort. The Journal of Pediatrics. 2004;**144**(2):169-176

[11] Hirai T et al. Brain structural changes in patients with chronic methylmercury poisoning in Minamata. Brain Research. 2023;**1805**:148278

[12] Guzzi G, La Porta CA. Molecular mechanisms triggered by mercury. Toxicology. 2008;**244**(1):1-12

[13] Jonnalagadda SB, Rao PV. Toxicity, bioavailability and metal speciation. Comparative Biochemistry and Physiology - Part C: Toxicology & Pharmacology. 1993;**106**(3):585-595

[14] Landrigan PJ et al. Human health and ocean pollution. Annals of Global Health. 2020;**86**(1):151

[15] Chen C. Methylmercury effects and exposures: Who is at risk? Environmental Health Perspectives. 2012;**120**(6):A224-A225

[16] Roman HA et al. Evaluation of the cardiovascular effects of methylmercury exposures: Current evidence supports development of a dose-response function for regulatory benefits analysis. Environmental Health Perspectives. 2011;**119**(5):607-614

[17] Stern AH. A review of the studies of the cardiovascular health effects of methylmercury with consideration of their suitability for risk assessment. Environmental Research. 2005;**98**(1):133-142

[18] Salonen JT et al. Intake of mercury from fish, lipid peroxidation, and the risk of myocardial infarction and coronary, cardiovascular, and any death in eastern Finnish men. Circulation. 1995;**91**(3):645-655

[19] Caballero B et al. Methylmercury-induced developmental toxicity is associated with oxidative stress and cofilin phosphorylation. Cellular and human studies. Neurotoxicology. 2017;**59**:197-209

[20] Joshi D et al. Reversal of methylmercury-induced oxidative stress, lipid peroxidation, and DNA damage by the treatment of N-acetyl cysteine: A protective approach. Journal of Environmental Pathology, Toxicology and Oncology. 2014;**33**(2):167-182

[21] Wang X et al. Low-dose methylmercury-induced apoptosis and mitochondrial DNA mutation in human embryonic neural progenitor cells. Oxidative Medicine and Cellular Longevity. 2016;**2016**:5137042

[22] Khan H et al. Mercury exposure induces cytoskeleton disruption and loss of renal function through epigenetic modulation of MMP9 expression. Toxicology. 2017;**386**:28-39

[23] Zuo K et al. L-ascorbic acid 2-phosphate attenuates methylmercury-induced apoptosis by inhibiting reactive oxygen species accumulation and DNA damage in human SH-SY5Y cells. Toxics. 2023;**11**(2):144

[24] Chistiakov DA et al. The role of mitochondrial dysfunction in cardiovascular disease: A brief review. Annals of Medicine. 2018;**50**(2):121-127

[25] Incalza MA et al. Oxidative stress and reactive oxygen species in endothelial dysfunction associated with cardiovascular and metabolic diseases. Vascular Pharmacology. 2018;**100**:1-19

[26] Migita K et al. Peroxynitrite-mediated matrix metalloproteinase-2 activation in human hepatic stellate cells. FEBS Letters. 2005;**579**(14):3119-3125

[27] Rajagopalan S et al. Reactive oxygen species produced by macrophage-derived foam cells regulate the activity of vascular matrix metalloproteinases in vitro. Implications for atherosclerotic plaque stability. The Journal of Clinical Investigation. 1996;**98**(11):2572-2579

[28] Viappiani S et al. Activation and modulation of 72kDa matrix metalloproteinase-2 by peroxynitrite and glutathione. Biochemical Pharmacology. 2009;**77**(5):826-834

[29] Hadler-Olsen E et al. Regulation of matrix metalloproteinase activity in health and disease. The FEBS Journal. 2011;**278**(1):28-45

[30] Springman EB et al. Multiple modes of activation of latent human fibroblast collagenase: Evidence for the role of a Cys73 active-site zinc complex in latency and a "cysteine switch" mechanism for activation. Proceedings of the National Academy of Sciences of the United States of America. 1990;**87**(1):364-368

[31] Bassiouni W, Ali MAM, Schulz R. Multifunctional intracellular matrix metalloproteinases: Implications in disease. The FEBS Journal. 2021;**288**(24):7162-7182

[32] Schulz CG et al. MMP-2 and MMP-9 and their tissue inhibitors in the plasma

of preterm and term neonates. Pediatric Research. 2004;**55**(5):794-801

[33] Baron MA et al. Matrix metalloproteinase 2 and 9 enzymatic activities are selectively increased in the myocardium of chronic Chagas disease cardiomyopathy patients: Role of TIMPs. Frontiers in Cellular and Infection Microbiology. 2022;**12**:836242

[34] Elahirad S et al. Association of matrix metalloproteinase-2 (MMP-2) and MMP-9 promoter polymorphisms, their serum levels, and activities with coronary artery calcification (CAC) in an Iranian population. Cardiovascular Toxicology. 2022;**22**(2):118-129

[35] Kurzepa J et al. The significance of matrix metalloproteinase (MMP)-2 and MMP-9 in the ischemic stroke. The International Journal of Neuroscience. 2014;**124**(10):707-716

[36] Jacob-Ferreira AL et al. Mercury exposure increases circulating net matrix metalloproteinase (MMP)-2 and MMP-9 activities. Basic & Clinical Pharmacology & Toxicology. 2009;**105**(4):281-288

[37] Edwards BA et al. Fifty years of volcanic mercury emission research: Knowledge gaps and future directions. Science of the Total Environment. 2021;**757**:143800

[38] Manceau A et al. Chemical forms of mercury in pyrite: Implications for predicting mercury releases in acid mine drainage settings. Environmental Science & Technology. 2018;**52**(18):10286-10296

[39] Manceau A et al. Biogenesis of mercury-sulfur nanoparticles in plant leaves from atmospheric gaseous mercury. Environmental Science & Technology. 2018;**52**(7):3935-3948

[40] Asaduzzaman A et al. Environmental mercury chemistry—in Silico. Accounts of Chemical Research. 2019;**52**(2):379-388

[41] Li Z et al. Reduction of mercury emissions from anthropogenic sources including coal combustion. Journal of Environmental Sciences (China). 2021;**100**:363-368

[42] Thanos Bourtsalas AC, Themelis NJ. Major sources of mercury emissions to the atmosphere: The U.S. case. Waste Management. 2019;**85**:90-94

[43] Sakamoto M, Nakamura M, Murata K. Mercury as a global pollutant and mercury exposure assessment and health effects. Nihon Eiseigaku Zasshi. 2018;**73**(3):258-264

[44] Li Y et al. Looping mercury cycle in global environmental-economic system modeling. Environmental Science & Technology. 2022;**56**(5):2861-2879

[45] Pavithra KG et al. Mercury sources, contaminations, mercury cycle, detection and treatment techniques: A review. Chemosphere. 2023;**312**(Pt 1):137314

[46] Song W et al. Effect of salinity and algae biomass on mercury cycling genes and bacterial communities in sediments under mercury contamination: Implications of the mercury cycle in arid regions. Environmental Pollution. 2021;**269**:116141

[47] Sharma Ghimire P et al. Microbial mercury methylation in the cryosphere: Progress and prospects. Science of the Total Environment. 2019;**697**:134150

[48] Trevors JT. Mercury methylation by bacteria. Journal of Basic Microbiology. 1986;**26**(8):499-504

[49] Baldi F. Microbial transformation of mercury species and their importance in the biogeochemical cycle of mercury.

Metal Ions in Biological Systems. 1997;**34**:213-257

[50] Wang J et al. Role of sulfur biogeochemical cycle in mercury methylation in estuarine sediments: A review. Journal of Hazardous Materials. 2022;**423**(Pt A):126964

[51] Luo J et al. Characterization of the trophic transfer and fate of methylmercury in the food web of Zhalong wetland, Northeastern China. Environmental Science and Pollution Research International. 2022;**29**(17):25222-25233

[52] Renzoni A, Zino F, Franchi E. Mercury levels along the food chain and risk for exposed populations. Environmental Research. 1998;**77**(2):68-72

[53] Schartup AT et al. A model for methylmercury uptake and trophic transfer by marine plankton. Environmental Science & Technology. 2018;**52**(2):654-662

[54] Castano A et al. Fish consumption patterns and hair mercury levels in children and their mothers in 17 EU countries. Environmental Research. 2015;**141**:58-68

[55] Connelly NA et al. Estimated exposure to mercury from fish consumption among women anglers of childbearing age in the Great Lakes region. Environmental Research. 2019;**171**:11-17

[56] Custodio FB et al. Total mercury in commercial fishes and estimation of Brazilian dietary exposure to methylmercury. Journal of Trace Elements in Medicine and Biology. 2020;**62**:126641

[57] Marrugo-Negrete J et al. Human health risk of methylmercury from fish consumption at the largest floodplain in Colombia. Environmental Research. 2020;**182**:109050

[58] Walker EV et al. Patterns of fish and whale consumption in relation to methylmercury in hair among residents of Western Canadian Arctic communities. BMC Public Health. 2020;**20**(1):1073

[59] Jo S et al. Estimation of the biological half-life of methylmercury using a population toxicokinetic model. International Journal of Environmental Research and Public Health. 2015;**12**(8):9054-9067

[60] Rand MD, Caito SW. Variation in the biological half-life of methylmercury in humans: Methods, measurements and meaning. Biochimica et Biophysica Acta - General Subjects. 2019;**1863**(12):129301

[61] Castoldi AF et al. Neurodevelopmental toxicity of methylmercury: Laboratory animal data and their contribution to human risk assessment. Regulatory Toxicology and Pharmacology. 2008;**51**(2):215-229

[62] Nielsen JB, Andersen O. The toxicokinetics of mercury in mice offspring after maternal exposure to methylmercury—effect of selenomethionine. Toxicology. 1992;**74**(2-3):233-241

[63] Ronconi-Kruger N et al. Methylmercury toxicity during heart development: A combined analysis of morphological and functional parameters. Cardiovascular Toxicology. 2022;**22**(12):962-970

[64] Ceccatelli S, Dare E, Moors M. Methylmercury-induced neurotoxicity and apoptosis. Chemico-Biological Interactions. 2010;**188**(2):301-308

[65] Bjorklund G et al. The toxicology of mercury: Current research and emerging trends. Environmental Research. 2017;**159**:545-554

[66] Farina M, Aschner M. Glutathione antioxidant system and methylmercury-induced neurotoxicity: An intriguing interplay. Biochimica et Biophysica Acta - General Subjects. 2019;**1863**(12):129285

[67] Unoki T et al. Spatio-temporal distribution of reactive sulfur species during methylmercury exposure in the rat brain. The Journal of Toxicological Sciences. 2022;**47**(1):31-37

[68] Yoshida E, Abiko Y, Kumagai Y. Glutathione adduct of methylmercury activates the Keap1-Nrf2 pathway in SH-SY5Y cells. Chemical Research in Toxicology. 2014;**27**(10):1780-1786

[69] Bridges CC, Zalups RK. Molecular and ionic mimicry and the transport of toxic metals. Toxicology and Applied Pharmacology. 2005;**204**(3):274-308

[70] Kanai Y, Endou H. Functional properties of multispecific amino acid transporters and their implications to transporter-mediated toxicity. The Journal of Toxicological Sciences. 2003;**28**(1):1-17

[71] Simmons-Willis TA et al. Transport of a neurotoxicant by molecular mimicry: The methylmercury-L-cysteine complex is a substrate for human L-type large neutral amino acid transporter (LAT) 1 and LAT2. The Biochemical Journal. 2002;**367**(Pt 1):239-246

[72] Yin Z et al. The methylmercury-L-cysteine conjugate is a substrate for the L-type large neutral amino acid transporter. Journal of Neurochemistry. 2008;**107**(4):1083-1090

[73] Takanezawa Y et al. Intracellular demethylation of methylmercury to inorganic mercury by organomercurial lyase (MerB) strengthens cytotoxicity. Toxicological Sciences. 2019;**170**(2):438-451

[74] Barlow NL, Bradberry SM. Investigation and monitoring of heavy metal poisoning. Journal of Clinical Pathology. 2023;**76**(2):82-97

[75] Kim JJ, Kim YS, Kumar V. Heavy metal toxicity: An update of chelating therapeutic strategies. Journal of Trace Elements in Medicine and Biology. 2019;**54**:226-231

[76] Sangvanich T et al. Novel oral detoxification of mercury, cadmium, and lead with thiol-modified nanoporous silica. ACS Applied Materials & Interfaces. 2014;**6**(8):5483-5493

[77] Al-Tikriti K, Al-Mufti AW. An outbreak of organomercury poisoning among Iraqi farmers. Bulletin of the World Health Organization. 1976;**53**(Suppl):15-21

[78] Greenwood MR. Methylmercury poisoning in Iraq. An epidemiological study of the 1971-1972 outbreak. Journal of Applied Toxicology. 1985;**5**(3):148-159

[79] Korbas M et al. The chemical nature of mercury in human brain following poisoning or environmental exposure. ACS Chemical Neuroscience. 2010;**1**(12):810-818

[80] Taber KH, Hurley RA. Mercury exposure: Effects across the lifespan. The Journal of Neuropsychiatry and Clinical Neurosciences. 2008;**20**(4):iv-389

[81] Zahir F, Rizwi SJ, Haq SK, Khan RH. Low dose mercury toxicity and human health. Environmental Toxicology and Pharmacology. 2005;**20**(2):351-360

[82] Amin-Zaki L et al. Intra-uterine methylmercury poisoning in Iraq. Pediatrics. 1974;**54**(5):587-595

[83] Eto K et al. A fetal type of Minamata disease. An autopsy case report with special reference to the nervous system. Molecular and Chemical Neuropathology. 1992;**16**(1-2):171-186

[84] Sakamoto M, Itai T, Murata K. Effects of prenatal methylmercury exposure: From Minamata disease to environmental health studies. Nihon Eiseigaku Zasshi. 2017;**72**(3):140-148

[85] Castoldi AF et al. Neurotoxicity and molecular effects of methylmercury. Brain Research Bulletin. 2001;**55**(2): 197-203

[86] Davidson PW et al. Effects of prenatal and postnatal methylmercury exposure from fish consumption on neurodevelopment: Outcomes at 66 months of age in the Seychelles child development study. JAMA. 1998;**280**(8):701-707

[87] Davidson PW et al. Longitudinal neurodevelopmental study of Seychellois children following in utero exposure to methylmercury from maternal fish ingestion: Outcomes at 19 and 29 months. Neurotoxicology. 1995;**16**(4):677-688

[88] Grandjean P, Herz KT. Methylmercury and brain development: Imprecision and underestimation of developmental neurotoxicity in humans. Mount Sinai Journal of Medicine. 2011;**78**(1):107-118

[89] Fujimura M, Usuki F. Cellular conditions responsible for methylmercury-mediated neurotoxicity. International Journal of Molecular Sciences. 2022;**23**(13):7218

[90] Bautista LE et al. Association of blood and hair mercury with blood pressure and vascular reactivity. WMJ. 2009;**108**(5):250-252

[91] Choi AL et al. Methylmercury exposure and adverse cardiovascular effects in Faroese whaling men. Environmental Health Perspectives. 2009;**117**(3):367-372

[92] Fillion M et al. A preliminary study of mercury exposure and blood pressure in the Brazilian Amazon. Environmental Health. 2006;**5**:29

[93] Salonen JT et al. Mercury accumulation and accelerated progression of carotid atherosclerosis: A population-based prospective 4-year follow-up study in men in Eastern Finland. Atherosclerosis. 2000;**148**(2):265-273

[94] Virtanen JK et al. Mercury, fish oils, and risk of acute coronary events and cardiovascular disease, coronary heart disease, and all-cause mortality in men in Eastern Finland. Arteriosclerosis, Thrombosis, and Vascular Biology. 2005;**25**(1):228-233

[95] Guallar E et al. Mercury, fish oils, and the risk of myocardial infarction. The New England Journal of Medicine. 2002;**347**(22):1747-1754

[96] Omanwar S et al. Modulation of vasodilator response via the nitric oxide pathway after acute methyl mercury chloride exposure in rats. BioMed Research International. 2013;**2013**:530603

[97] Grotto D et al. Low level and sub-chronic exposure to methylmercury induces hypertension in rats: Nitric oxide depletion and oxidative damage as possible mechanisms. Archives of Toxicology. 2009;**83**(7):653-662

[98] Roque CR et al. Methylmercury chronic exposure affects the expression of DNA single-strand break repair genes, induces oxidative stress, and chromosomal abnormalities in young

dyslipidemic APOE knockout mice. Toxicology. 2021;**464**:152992

[99] Santos Ruybal MCP et al. Methylmercury poisoning induces cardiac electrical remodeling and increases arrhythmia susceptibility and mortality. International Journal of Molecular Sciences. 2020;**21**(10):3490

[100] Silva JL et al. Oral methylmercury intoxication aggravates cardiovascular risk factors and accelerates atherosclerosis lesion development in ApoE knockout and C57BL/6 mice. Toxicology Research. 2021;**37**(3):311-321

[101] Sumathi T, Jacob S, Gopalakrishnan R. Methylmercury exposure develops atherosclerotic risk factors in the aorta and programmed cell death in the cerebellum: Ameliorative action of Celastrus paniculatus ethanolic extract in male Wistar rats. Environmental Science and Pollution Research International. 2018;**25**(30):30212-30223

[102] Hagele TJ et al. Mercury activates vascular endothelial cell phospholipase D through thiols and oxidative stress. International Journal of Toxicology. 2007;**26**(1):57-69

[103] Kishimoto T, Oguri T, Tada M. Effect of methylmercury (CH_3HgCl) injury on nitric oxide synthase (NOS) activity in cultured human umbilical vascular endothelial cells. Toxicology. 1995;**103**(1):1-7

[104] Mazerik JN et al. Phospholipase A2 activation regulates cytotoxicity of methylmercury in vascular endothelial cells. International Journal of Toxicology. 2007;**26**(6):553-569

[105] Ghizoni H et al. Superoxide anion generation and oxidative stress in methylmercury-induced endothelial toxicity in vitro. Toxicology In Vitro. 2017;**38**:19-26

[106] Franco JL et al. Methylmercury neurotoxicity is associated with inhibition of the antioxidant enzyme glutathione peroxidase. Free Radical Biology & Medicine. 2009;**47**(4): 449-457

[107] Robitaille S, Mailloux RJ, Chan HM. Methylmercury alters glutathione homeostasis by inhibiting glutaredoxin 1 and enhancing glutathione biosynthesis in cultured human astrocytoma cells. Toxicology Letters. 2016;**256**:1-10

[108] Lian M et al. Assessing oxidative stress in Steller Sea lions (Eumetopias jubatus): Associations with mercury and selenium concentrations. Comparative Biochemistry and Physiology Part C: Toxicology & Pharmacology. 2020;**235**:108786

[109] Mori N, Yasutake A, Hirayama K. Comparative study of activities in reactive oxygen species production/ defense system in mitochondria of rat brain and liver, and their susceptibility to methylmercury toxicity. Archives of Toxicology. 2007;**81**(11):769-776

[110] Bonetti PO, Lerman LO, Lerman A. Endothelial dysfunction: A marker of atherosclerotic risk. Arteriosclerosis, Thrombosis, and Vascular Biology. 2003;**23**(2):168-175

[111] Hadi HA, Carr CS, Al Suwaidi J. Endothelial dysfunction: Cardiovascular risk factors, therapy, and outcome. Vascular Health and Risk Management. 2005;**1**(3):183-198

[112] Sun HJ et al. Role of endothelial dysfunction in cardiovascular diseases: The link between inflammation and hydrogen sulfide. Frontiers in Pharmacology. 2019;**10**:1568

[113] Thomas DD et al. The chemical biology of nitric oxide: Implications in cellular signaling. Free Radical Biology & Medicine. 2008;**45**(1):18-31

[114] Costa TJ et al. The homeostatic role of hydrogen peroxide, superoxide anion and nitric oxide in the vasculature. Free Radical Biology & Medicine. 2020;**162**:615-635

[115] Shinyashiki M et al. Selective inhibition of the mouse brain Mn-SOD by methylmercury. Environmental Toxicology and Pharmacology. 1996;**2**(4):359-366

[116] Shinyashiki M et al. Differential changes in rat brain nitric oxide synthase in vivo and in vitro by methylmercury. Brain Research. 1998;**798**(1-2):147-155

[117] Nagase H, Visse R, Murphy G. Structure and function of matrix metalloproteinases and TIMPs. Cardiovascular Research. 2006;**69**(3):562-573

[118] Van Wart HE, Birkedal-Hansen H. The cysteine switch: A principle of regulation of metalloproteinase activity with potential applicability to the entire matrix metalloproteinase gene family. Proceedings of the National Academy of Sciences of the United States of America. 1990;**87**(14):5578-5582

[119] Jacob-Ferreira AL et al. A common matrix metalloproteinase (MMP)-2 polymorphism affects plasma MMP-2 levels in subjects environmentally exposed to mercury. Science of the Total Environment. 2011;**409**(20):4242-4246

[120] Jacob-Ferreira AL et al. A functional matrix metalloproteinase (MMP)-9 polymorphism modifies plasma MMP-9 levels in subjects environmentally exposed to mercury. Science of the Total Environment. 2010;**408**(19):4085-4092

[121] Belo VA et al. Assessment of matrix metalloproteinase (MMP)-2, MMP-8, MMP-9, and their inhibitors, the tissue inhibitors of metalloproteinase (TIMP)-1 and TIMP-2 in obese children and adolescents. Clinical Biochemistry. 2009;**42**(10-11):984-990

[122] Palei AC et al. Comparative assessment of matrix metalloproteinase (MMP)-2 and MMP-9, and their inhibitors, tissue inhibitors of metalloproteinase (TIMP)-1 and TIMP-2 in preeclampsia and gestational hypertension. Clinical Biochemistry. 2008;**41**(10-11):875-880

[123] Ali MA et al. Titin is a target of matrix metalloproteinase-2: Implications in myocardial ischemia/reperfusion injury. Circulation. 2010;**122**(20):2039-2047

[124] Sawicki G et al. Degradation of myosin light chain in isolated rat hearts subjected to ischemia-reperfusion injury: A new intracellular target for matrix metalloproteinase-2. Circulation. 2005;**112**(4):544-552

[125] Sung MM et al. Matrix metalloproteinase-2 degrades the cytoskeletal protein alpha-actinin in peroxynitrite mediated myocardial injury. Journal of Molecular and Cellular Cardiology. 2007;**43**(4):429-436

[126] Wang W et al. Intracellular action of matrix metalloproteinase-2 accounts for acute myocardial ischemia and reperfusion injury. Circulation. 2002;**106**(12):1543-1549

[127] Brown RD et al. Cytokines regulate matrix metalloproteinases and migration in cardiac fibroblasts. Biochemical and Biophysical Research Communications. 2007;**362**(1):200-205

[128] McQuibban GA et al. Matrix metalloproteinase activity inactivates

the CXC chemokine stromal cell-derived factor-1. The Journal of Biological Chemistry. 2001;**276**(47):43503-43508

[129] Van Den Steen PE et al. Gelatinase B/MMP-9 and neutrophil collagenase/MMP-8 process the chemokines human GCP-2/CXCL6, ENA-78/CXCL5 and mouse GCP-2/LIX and modulate their physiological activities. European Journal of Biochemistry. 2003;**270**(18):3739-3749

[130] Rizzi E et al. Temporal changes in cardiac matrix metalloproteinase activity, oxidative stress, and TGF-beta in renovascular hypertension-induced cardiac hypertrophy. Experimental and Molecular Pathology. 2013;**94**(1):1-9

[131] Wang Y et al. Matrix metalloproteinase-9 induces cardiac fibroblast migration, collagen and cytokine secretion: Inhibition by salvianolic acid B from salvia miltiorrhiza. Phytomedicine. 2011;**19**(1):13-19

[132] Chandra S et al. Epigenetics and expression of key genes associated with cardiac fibrosis: NLRP3, MMP2, MMP9, CCN2/CTGF and AGT. Epigenomics. 2021;**13**(3):219-234

[133] Fan D et al. Cardiac fibroblasts, fibrosis and extracellular matrix remodeling in heart disease. Fibrogenesis & Tissue Repair. 2012;**5**(1):15

[134] Halbach S, Schonsteiner G, Vierling W. The action of organic mercury compounds on the function of isolated mammalian heart muscle. European Journal of Pharmacology. 1989;**167**(2):255-264

[135] Massaroni L et al. Effects of mercury on the mechanical and electrical activity of the Langendorff-perfused rat heart. Brazilian Journal of Medical and Biological Research. 1992;**25**(8):861-864

[136] Su JY, Chen W. The effect of methylmercury on isolated cardiac tissues. The American Journal of Pathology. 1979;**95**(3):753-764

[137] Dorea JG et al. Hair mercury (signature of fish consumption) and cardiovascular risk in munduruku and kayabi Indians of Amazonia. Environmental Research. 2005;**97**(2):209-219

[138] Yoshizawa K et al. Mercury and the risk of coronary heart disease in men. The New England Journal of Medicine. 2002;**347**(22):1755-1760

[139] Yang L et al. Methyl mercury suppresses the formation of the tail primordium in developing zebrafish embryos. Toxicological Sciences. 2010;**115**(2):379-390

[140] Fu X, Yang X, Cui Q. Deciphering the possible role of H2O2 in methylmercury-induced neurotoxicity in Xenopus laevis. Molecular & Cellular Toxicology. 2020;**16**:301-309

Chapter 6

Oxidative Stress and Reproduction Health: Physiology, Pathology, and Clinical Biomarkers

Ramona D'Amico, Salvatore Cuzzocrea, Marika Cordaro and Rosanna Di Paola

Abstract

Reactive oxygen species (ROS) are free radicals derived from oxygen during normal cellular metabolism. Cells, under aerobic conditions, have a defense system against ROS, and in normal circumstances, there is an appropriate balance between prooxidants and antioxidants. When an overproduction of ROS develops or the body fails to eliminate ROS in excess, oxidative stress arises, during which ROS accumulate and damage cells and tissues. ROS plays a crucial role in the physiological processes and signaling pathways associated in both male and female fertility. In females, oxidative stress acts as a mediator in the modulation of important ovarian functions, and its complications such as abortions, recurrent pregnancy losses, preeclampsia, and gestational diabetes. In males, ROS plays an important role in normal physiological processes such as capacitation, hyperactivation, acrosome reaction, and sperm-oocyte fusion to ensure appropriate fertilization. However, high concentrations of ROS cause sperm pathologies (ATP depletion) in the form of inadequate axonemal phosphorylation or lipid peroxidation, resulting in a loss of sperm motility and viability. This chapter will highlight the mechanisms, production, physiological, and pathophysiological roles of ROS in relation to the male and female reproductive system, and recent advances in diagnostic methods that use ROS as biochemical markers.

Keywords: oxidative stress, biomarkers, physiology, pathology, reproduction health, male, female

1. Introduction

The key criteria for successful fertilization are the formation of mature and normal spermatozoa from maturing spermatids. After immature germ cells are produced in the testis's seminiferous tubules, they travel to the epididymis, where they are stored and mature under the influence of the neighboring epithelial cells. This epididymal maturation phase converts nonmotile spermatozoa into motile germ cells capable of fertilizing an oocyte [1]. ROS formation is a feature shared by all cells, including spermatozoa, in all mammalian species. There are two hypothesized pathways for ROS generation by spermatozoa: (a) *via* the sperm membrane's nicotinamide adenine dinucleotide phosphate (NADPH) oxidase system and/or (b) *via* the sperm's nicotinamide adenine dinucleotide-dependent redox reaction [2]. The second significant source of ROS detected in sperm is leukocytes, which operate as a vital element of the cellular defense mechanism against infection, varicocele, spinal cord damage, chronic sexual absentia, and inflammation. To combat infectious pathogens, leukocytic infiltration is increased during infection. Increased inflammatory cytokine production, such as Interleukin-8 (IL-8), combined with reduced SOD synthesis results in a respiratory burst and increased ROS generation, leading to oxidative stress [3]. Sertoli cells have also been shown to be capable of producing ROS [4]. Scavestrogens (synthetic steroidal estrogens with significant antioxidant capabilities) have been proven in studies to reduce the generation of ROS by Sertoli cells. *In vitro*, scavestrogens may scavenge free radical scavengers and limit iron-induced cell damage [5]. Thus, it is possible that Sertoli cells, under normal *in vivo* circumstances, assist spermatogenesis *via* ROS generation [6]. Mammalian sperm must have features in order to operate properly and reproduce, including correct morphology, motility, and the capacity to undertake certain processes such as capacitation and acrosomal response. According to research, physiological amounts of ROS operate as intracellular signaling molecules required for several physiological processes such as maturation, hyperactivation, capacitation, acrosomal response, and oocyte-sperm union [7]. ROS activities are mostly carried out through regulating the redox state of cysteine residues. ROS not only enhances protein tyrosine kinase activity but it also suppresses phosphotyrosine activity, which typically dephosphorylates tyrosine. Tyrosine in the fibrous sheath enclosing the axoneme of the sperm flagellum is phosphorylated by protein tyrosine kinase. Capacitation results in increased phosphorylation of tyrosine. A-kinase anchoring proteins are phosphorylated proteins that bind PKA to the fibrous sheath of the sperm, suggesting a possible function in sperm hyperactivity. Spermatozoa that emerge from the testis are not fully developed, and hence cannot fertilize an egg. In addition to storage and transportation, the epididymis is important in spermatozoa obtaining the capacity to fertilize. In the epididymal environment, spermatozoa undergo changes such as progressive motility, chromatin condensation, and plasma membrane remodeling, all of which are required for capacitation and fertilization in the female reproductive tract. Despite the fact that spermatozoa are extremely vulnerable to oxidative stress-mediated destruction, normal amounts of ROS are required for epididymal sperm maturation [8]. Several antioxidants help to maintain a physiological redox equilibrium in the epididymis. As a result, in addition to functioning as ROS scavengers, antioxidants control the bioavailability of oxidants, which are then employed to catalyze protamine and flagellar sulfoxidation. While the testis partially completes disulfide production on cysteine sulfhydryls of sperm protamines, the majority of sulfoxidation occurs in the epididymal environment [9]. During capacitation, spermatozoa undergo molecular changes that guarantee the continuation of the acrosome reaction and fusion with the egg membrane. These include (a) basification of intracellular pH, (b) activation of cAMP-dependent pathways, (c) removal of cholesterol from the sperm membrane, and (d) protein phosphorylation by cAMP-dependent kinases at serine, threonine, and tyrosine residues. ROS are essential regulators of these activities [10]. Suppression of capacitation has been reported in epididymal spermatozoa incubated with catalase, which inhibits H_2O_2 synthesis. Several studies have highlighted the significance of ROS in regulating the cAMP pathway, which includes PKA activation and phosphorylation of PKA substrates. Indeed, elevated amounts of the second messengers cAMP and Ca^{2+}, as well as a high rate of protein phosphorylation, have been linked to ROS generation during the capacitation process [11]. The oxidation of cholesterol and subsequent efflux of cholesterol from the plasma membrane is required for the preparation of spermatozoa for the acrosome reaction [12]. The basic condition for successful fertilization is sperm-oocyte fusion, which is followed by sperm head penetration through the zona pellucida. To do the same, hyperactivated spermatozoa must overcome the layer of cumulus cells to reach the oocyte's zona pellucida. The completion of an acrosome reaction (AR) refers to the conclusion of the spermatozoon's maturational phases at which time it obtains the final fertilizing capacity. ROS may aid the acrosome process by inducing phosphorylation of certain plasma membrane proteins [5]. The AR can be induced by both physiological and nonphysiological stimuli such as ROS, the zona pellucida, or progesterone [13]. The large amount of membrane fatty acids released during AR increases the sperm plasma membrane fluidity essential for fusion with the oocyte [14]. ROS plays an important role in enhancing plasma membrane fluidity during sperm-oocyte fusion by mediating the activation of metabolic pathways for capacitation, which is followed by a successful acrosome response. ROS suppresses PLA2 deactivation by inhibiting protein tyrosine phosphatase activity throughout spermatozoa capacitation. Thus, active PLA2 may cleave secondary fatty acids from membrane phospholipid triglycerides, increasing plasma membrane fluidity [15].

2. ROS physiology in women

OS plays an important role in a variety of physiological and pathological processes of the female reproductive system. As ROS influences many physiological and pathological processes in the ovaries, as well as the peritoneal environment, there are methods through which OS may impair female fertility. The main sources of ROS in the follicular fluid microenvironment include cytokines, macrophages, and leukocytes. Because ROS are regarded key inducers of ovulation, a particular level of ROS is required [16]. However, one way that OS may impair female fertility is by the action of ROS on female germ cells. The fundamental mechanism by which the postovulatory oocytes lose their developmental ability after ovulation appears to be OS. Postovulatory oocytes enter apoptosis and lose functioning as a result of a complicated chain of events caused by an increase in oxidative stress [14]. One common marker of postovulatory oocyte aging is zona pellucida induration, which can be caused by OS exposure [14]. Ovoperoxidase, which is present in the cortical granules that are ejected from the oocyte's surface during an exocytotic process, is fueled by OS and aids in oocyte aging. Furthermore, excessive ROS production might damage oocyte DNA, resulting in incorrect fertilization [14]. According to research, psychological stress and aging may produce oxidative imbalance [17]. Furthermore, oxidative stress may impair granulosa cells' capacity to produce steroid hormones such as follicle-stimulating hormone (FSH) and estradiol (E2), potentially affecting oocyte quality [18]. The increased OS in the granulosa cells, which is coupled with a reduction in the expression of the follicle-stimulating hormone receptor (FSHR) and a dysregulation of the FSHR signaling pathway, may be linked to the poor response to FSH and impaired steroidogenic activity in older women [19]. O_2 and its scavenging molecules have been extensively studied in terms of their physiologic and pathologic effects. The presence of O_2 and SOD antioxidant enzymes in the mammalian female reproductive system has been found in the uterus [20]. Throughout the peri-implantation stage, SOD levels and activity rise, ostensibly to protect the embryo throughout the implantation process. Copper, zinc SOD isoenzyme, and SOD activity have also been found in rabbit and human fallopian tube epithelium and oviduct fluid [21]. During the estrous cycle, O_2 and SOD levels in the rat ovary alter inversely. Superoxide dismutase activity and the copper, zinc SOD isoenzyme is seen in developing follicles, granulosa cell membranes in Graafian follicles, and postovulatory follicles [22]. Further research on rat and human ovaries suggests that O_2 and SOD enzymes may play a function in the ovulation process and oocyte development, and current results reveal that O_2 and its scavenging molecules have a regulatory role in progesterone generation by the mammalian corpus luteum [23]. The oocyte maturation process needs a steady supply of energy in the form of adenosine triphosphate (ATP) to feed the transcription process and expand the size of the follicles, as well as the oocyte. The content of ATP during metaphase II (MII) arrest in the human egg is positively connected with successful fertilization and IVF result. During the maturation process, the mitochondrial electron transport chain generates ATP, which leads in the creation of ROS. This high quantity of ROS is oxidized by antioxidant enzymes found in the follicular fluid, including as catalase, SOD, glutathione transferase (GST), heat shock protein, and protein isomerase, which finally protect the oocytes from the detrimental effects of ROS [24]. Aside from mitochondrial respiration and oxidative phosphorylation, various other processes contribute to ROS emission in follicles. The corpus luteum is signaled by luteinizing hormone (LH) to generate and release progesterone. The procedure is quite complex and is controlled at various stages, which are detailed below. Following LH signaling, cholesterol is first transformed to pregnenolone in the mitochondria by cytochrome P450 side-chain breakage, which produces ROS [25]. Pregnenolone is further converted to progesterone in the endoplasmic reticulum by 3-β-hydroxysteroid dehydrogenase, which needs oxidized nicotinamide adenine dinucleotide (NAD^+) as the hydrogen acceptor. NAD^+ is produced by the oxidation of NADH by ascorbic radicals *via* a free radical process mediated by ROS [26]. This demonstrates the critical function that ROS plays in the process of LH-induced progesterone release. According to one study, LH increases the intraluteal expression of copper/zinc-dependent SOD (Cu/Zn-SOD), manganese-dependent SOD (Mn-SOD), and catalase to preserve luteal cell viability and ROS equilibrium. According to this study, LH not only stimulates ROS generation but also controls it by promoting the expression of antioxidant enzymes, hence preserving the redox state [27]. In rat preovulatory granulosa cells, LH increases P4 synthesis in part by activating P450scc, which generates ROS. It has been demonstrated that ROS cause the apoptosis of bovine luteal cells. As a result, pulsatile LH secretion appears to both boost P4 and cause the formation of luteolytic ROS throughout the luteal phase. Moreover, P4 production in the bovine CL peaks around the middle of the luteal phase, which is also when ROS generation in the CL may be at its maximum. Due to LH's stimulation of antioxidant enzyme synthesis, CL function may be preserved at the mid luteal phase. The overall findings of that investigation show that LH increases the viability of luteal cells by promoting intraluteal expressions of and by promoting SOD activity to preserve luteal function [27, 28].

3. ROS pathology in male

Leukocytospermia, a condition that is closely associated to problems in sperm function, is characterized by the presence of more than a million peroxidase-positive leukocytes per milliliter of semen [29]. The physiological process of excess cytoplasm being extruded during sperm maturation is common. However, when spermiogenesis is disrupted, spermatozoa retain excess cytoplasm around the midpiece, which impairs the midpiece's capabilities. Most immature spermatozoa with a distorted head shape and cytoplasmic retention produce seminal ROS [30]. The accumulation of metabolic enzymes, such as glucose-6-phosphate dehydrogenase (G6PD) and NADPH oxidase, which are directly implicated in the synthesis of free radicals through the intermediate NADPH creation, is brought on by the preservation of extra cytoplasm in immature sperm [30]. Both NADPH oxidase in the plasma membrane and NADH-dependent oxidoreductase (diaphorase) in the mitochondria allow normal spermatozoa to produce ROS [31]. Thus, it is possible to infer that Sertoli cells support spermatogenesis through ROS generation under normal *in vivo* settings. To fully understand Sertoli cells' contribution to the creation of ROS, more research is necessary. One of the main reasons of male subfertility and infertility is varicocele, an aberrant venous dilatation in the pampiniform plexus around the spermatic cord that affects roughly 40% of male partners in overall infertile couples [32]. The pathophysiology of changes in sperm functions brought on by varicocele is explained by a number of processes. However, it has been shown that the most frequent mechanism for OS-induced sperm dysfunctions involves varicocele-mediated testicular heat and hypoxia [33]. In comparison to healthy fertile donors, oxidative stress indices, including ROS and lipid peroxidation, are more common in semen samples from varicocele-affected infertile individuals, according to a meta-analysis [34]. The degree of varicocele has been suggested to directly correlate with seminal ROS levels [35]. The inflammatory condition affecting the male genital system that has received the most research to date is prostatitis. Depending on the pathogen strain, pathogens can directly impact sperm in cases of acute and chronic bacterial prostatitis or indirectly through the activation of cytokines such as IL6, IL8, or tumor necrosis factor (TNF) [36]. OS is caused by elevated cytokine levels, which can also influence spermatozoa [37] and could cause a systemic reaction by reducing testosterone hormone levels. Elevated IL6, IL8, and TNF-levels may worsen sperm transit during ejaculation when the infection spreads to the testis [38]. The paternal genetic contribution to the embryo is typically decreased by increased OS, which frequently has a direct impact on sperm DNA [39]. Leukocytes are stimulated by prostate infection and inflammation and can generate ROS. Because a high level of ROS can affect up to 35% of men seeking infertility treatment, infections must be treated very away [40]. This is mostly as a result of the possibility of oligozoospermia, azoospermia, or asthenozoospermia as a result of untreated prostatitis [41]. Infections can also lead to chronic illnesses, which are more challenging to cure than acute illnesses. For the treatment of bacterial prostatitis, fluoroquinolones, tetracyclines, macrolides, and trimethoprim (singly or in combination with sulfamethoxazole) were all specifically mentioned [42]. The mainstay of treatment for bacterial prostatitis continues to be antibiotic medication, and numerous studies have shown that antibiotics can considerably improve sperm characteristics and conception rates. However, given that recent research has concentrated on multidrug resistance, antibiotic treatment must be administered with caution and after the proper drug resistance testing [43]. When a patient's first-line therapy is ineffective, combination therapy is frequently recommended [44]. Ciprofloxacin and rifampin appear to be effective against Staphylococcus aureus, and fluoroquinolones bind with metal cations such as aluminum, magnesium, calcium, iron, and zinc, while the majority of combinations are ineffective or additive. The serum medication concentrations that are accessible for tissue penetration are significantly reduced by this pharmacological interaction [45]. The amount of zinc produced by prostate epithelial cells can be three times that of the majority of other mammalian cells [46]. Despite what was just said, zinc buildup in the prostate may prevent fluoroquinolones from working to their full potential. Due to the direct effect of colonized bacteria or other pathogens on spermatozoa when they travel through the urethra during ejaculation, infection or inflammation of the urethra may result in infertility. By reducing the volume and/or the number of sperm, tissue scarring as a result of infection may prevent sperm from depositing in the female reproductive tract [47]. A genitourinary condition known as urethritis is frequently seen in males. This inflammatory disease has been associated with a number of infections, including Mycoplasma genitalium (MG), Neisseria gonorrhoeae (GC), Trichomonas vaginalis (TV), Chlamydia trachomatis (CT), Ureaplasma urealyticum (UU), Herpes simplex virus (HSV), and adenovirus [48]. The Staphylococcus with coagulase-negative. In women, Staphylococcus saprophyticus frequently causes urinary tract infections. However, it has been demonstrated to cause urethral infection in males, albeit its relevance in this context is uncertain [49]. Numerous problems have been connected to it, including acute epididymitis, orchitis, and prostatitis. Male infertility has an unclear etiology in roughly 50% of cases; however, it is evident that 30–80% of infertile men have a high level of ROS in their ejaculate. The term "male oxidative stress infertility (MOSI)" was created by Agarwal et al. to refer to male infertility brought on by OS because of the close association between OS and male infertility [50]. Leukocytes in the seminal fluid and immature sperm with a morphologically defective head and cytoplasmic retention are the two main sources of endogenous ROS in human semen [51]. Extrinsic ROS are created during male genital tract infection, and leukocyte chemotaxis and activation encourage further inflammatory responses. Leukocytes activate the myeloperoxidase system, which generates ROS, to fight infections [52]. The overproduction of ROS by leukocytes can result in OS in the seminal fluid. However, the origins of intrinsic ROS are defective and immature spermatozoa. It has been demonstrated to have harmful effects on sperm, including sperm DNA fragmentation (SDF), LPO, and apoptosis in germ cells. SDF may be caused by increased ROS generation and inadequate antioxidant protection in sperm [53]. Through the activation of sperm caspases and the generation of endonuclease, OS has the capacity to harm sperm DNA either directly or indirectly. The failure of the chromatin structure substitution from histone to protamine during the process of spermiogenesis is thought to be the cause of SDF since it results in DNA vulnerability. The migration of spermatozoa from the seminiferous tubules to the cauda epididymis *via* the rete testis causes excessive ROS exposure, which has been reported to cause DNA damage [54]. This reaction results in the formation of 8-OH-guanine and 8-OH-20-deoxyguanosine (8-OHdG), an oxidized guanine adduct. In the lab, increased levels of 8-OHdG have been connected to DNA fragmentation and strand breaks [55]. DNA fragmentation can happen to both single- and double-stranded (ds-) DNA [56]. DNA repair is only possible during specific stages of the spermiogenesis process, and it ceases to be active during the nuclear condensation of the epididymis. The human oocyte represents the next opportunity for ss-DNA break repair; however, SDF repair efficiency decreases with increasing maternal age [57]. A break in the ds-DNA causes genomic instability and ultimately causes cellular death when DNA repair does not take place [58]. The "late paternal effect" is hypothesized to be induced by the presence of unrepaired SDF above a specific threshold, which has a negative impact on embryo development and pregnancy outcomes [59]. A cleavage-stage embryo experiences a major activation of embryonic genome expression on the second day of human embryo development (4 cells), and the dependence of embryogenesis on maternal factors is replaced by the dependence on the embryo's own genome [60]. As a result, after fertilization, SDF in a spermatozoon may have a deleterious impact on the outcomes of blastulation, implantation, and pregnancy. The "early paternal effect" [61] was the term coined by the researchers to describe how OS negatively affects the growth of cleavage embryos. The effect of SDF on the results of ART has been the subject of numerous investigations [62]. A meta-analysis found a positive correlation between SDF and miscarriage and an inverse correlation between SDF with pregnancy outcome [62]. SDF is one of the elements that might lead to repeated pregnancy losses, thus accurate assessment and control could help couples with their issue. Apoptosis is acknowledged as a biologically designed cell death due to the DNA fragmentation that is brought on by a number of cell death signaling and regulatory mechanisms [63]. Ds-DNA breakage brought on by ROS may result in apoptosis. Additionally, ROS leads to mitochondrial membrane rupture, which releases the signaling molecule cytochrome-C, which can then activate apoptotic caspases and annexin-V phosphatidylserine-binding activity [64]. High amounts of cytochrome-C seen in the seminal plasma of infertile men may cause severe mitochondrial damage [65].

4. ROS pathology in female

The vascular endothelium-produced ROS may contribute to regular cellular signaling processes. They might also have a significant role in the development of endothelial dysfunction. Endometrial tissue can be found outside the uterus in endometriosis, a benign, estrogen-dependent, and persistent gynecological condition. According to a report, erythrocytes and apoptotic endometrioma cells produce high ROS that cause OS, as do activated macrophages that are brought in to phagocytize apoptotic cells [66]. In addition, endometriosis-afflicted women produce higher levels of the ROS-producing enzyme xanthine oxidase, which is thought to be another cause of excessive ROS [67]. Additionally, OS causes local inflammation that raises cytokine levels, as well as other variables that support endometriosis [68]. Phagocytic cells, which are the primary sources of ROS and RNS, are recruited and activated by pro-inflammatory and chemotactic cytokines. Endometriosis patients' peritoneal macrophages and endometriotic lesions have been found to be activated by OS [69]. Mitogen-activated protein kinase (MAPK) and extracellular regulated kinase (ERK1/2) are activated in endometriotic cells in a manner similar to tumor cells by elevated ROS and consequent cellular proliferation [70]. More gravely, the rise in ROS in endometriosis patients can have a negative impact on the development of the fetus, causing fetal dysmorphogenesis, IUGR, or spontaneous miscarriage [71]. A vascular pregnancy disease called preeclampsia frequently involves poor placental growth. It is a multisystem illness that can affect women with normal blood pressure. Because OS increases p38 MAPK nitration, which lowers its catalytic activity, it can result in the poor implantation and growth restriction seen in preeclampsia. The increased levels of MDA, a marker of lipid peroxidation, in preeclampsia patients have been used to demonstrate elevated ROS concentrations [72]. Under normal circumstances, preeclampsia's vascular endothelial dysfunction is the main factor contributing to the disruption of circulatory homeostasis. Low anticoagulant activity and a propensity to promote vasoconstriction define it. The endothelial dysfunction related to preeclampsia appears to be significantly influenced by ROS [73]. Injury to the vascular endothelium caused by increased placental ROS or impaired antioxidant activity [74] is the pathogenic event in preeclampsia. The rise of ROS can be attributed to a variety of factors. A prominent source of ROS, for instance, is neutrophil regulation in preeclampsia, which increases the production of the SO anion and decreases NO levels, ultimately damaging endothelial cells in preeclampsia patients. Preeclampsia is associated with elevated levels of TNF and oxLDL, which have been found to activate the endothelium isoform of NAD(P)H oxidase and consequently raise SO anion levels. According to these findings, consuming antioxidants to combat increased lipid peroxidation may harm the vascular endothelium and contribute to the pathophysiology of preeclampsia [75]. In comparison to women without the condition, preeclamptic women create more ROS and have higher NAD(P)H expression [76]. More particular, it has been noted that early-onset preeclampsia causes women to create more SO anion than late-onset preeclampsia does [77]. Affected women also have reduced levels of vitamins C and E and a lowered total antioxidant status (TAS), as well as placental GPx [78]. Preeclampsia seems to be more likely in people who do not consume enough vitamin C, and some research suggests that multivitamin supplementation during pregnancy can reduce preeclampsia risk in women who are normal weight or underweight [79]. Studies on the effects of restraint stress on uterine and embryo implantation in pregnant mice have been conducted. These studies focused on uterine histomorphology study and changes to the uterine local microenvironment. The mice were treated to restraint stress beginning on embryonic day 1 (E1), according to Liu Guanhui et al. This study showed that restraint stress raised corticosterone (CORT) levels in plasma and dramatically increased uterine natural killer (uNK) cells in the endometrium, along with a decrease in the density of mast cells in the myometrium. In addition, the $CD3^+CD4^+/CD3^+CD8^+$ T cell ratio was significantly reduced by constraint stress. Additionally, the content of MDA rose, and antioxidant capacity was weakened [41]. Restraint stress also decreased the weight of the uterus, ovary, and food consumption with weight loss, while it also decreased the relative endometrial area and uterine gland area. Restraint stress also reduced VEGF expression and micro-vessel density [80]. During reproductive life, only a tiny percentage of the ovarian follicles present at birth mature to ovulation, while the remainder undergo a degenerative process known as "atresia." Follicular atresia is caused by granulosa cell death, according to research [81]. Excessive ROS production causes antral follicular atresia by triggering granulosa cell death, according to mounting data. ROS are also implicated in the decrease of granulosa cell sensitivity to gonadotrophin hormones and the loss of steroidogenic activity, both of which are symptoms of follicular atresia. Furthermore, ROS concentrations rose throughout the regression phase in the rat ovarian corpus luteum. Thus, ROS might initiate apoptosis in the luteal cells and inhibit their function at an appropriate time during the female menstrual cycle [82].

5. ROS as biomarkers

Biomarkers of oxidative stress (OS) can be classified as molecules that are modified by interactions with ROS in the microenvironment, and molecules of the antioxidant system that change in response to increased redox stress. DNA, lipids (including phospholipids), proteins, and carbohydrates are examples of molecules that can be modified by excessive ROS *in vivo*. There are several methods to measure ROS in the laboratory setting, which can be classified into direct and indirect assays [83]. The indirect measurements of ROS include enzymatic antioxidant levels such as SOD, CAT, GPx, and reduced glutathione (GSH) *via* means of spectrophotometric measurement. As regards the direct assays of the ROS, these include the direct measurement of total antioxidant capacity (TAC), detection of ROS species *via* chemiluminescence, and the use of florescent markers for ROS, which can be measured using fluorescence microscopy and flow cytometry [83]. Some of the most widely used methods to detect ROS levels in both male and female reproductive systems are described below.

5.1 ROS measurement by chemiluminescence assay

Chemiluminescence measures the light that is emitted in a reaction when reagents are added to a biological sample [83]. Luminol and Lucigen are two probes that are used in chemiluminescent assays to detect ROS [84]. Lucigenin is best suited to detect $O_2^{-\bullet}$ as it is positively charged, which renders it membrane-impermeable and allows it to react with $O_2^{-\bullet}$ in the extracellular space [85]. Unlike Lucigen, Luminol is uncharged and is, therefore, membrane permeable; this allows it to react with ROS in both the intra- and extracellular spaces. Luminol reacts with a variety of ROS, including $O_2^{-\bullet}$, H_2O_2, and hydroxyl radicals ($OH^\bullet$) [86]. This probe, however, is unable to differentiate between the types of ROS and therefore measures global ROS [86].

5.2 ROS measurement by flow cytometry

Flow cytometry involves the use of florescent markers to measure ROS and RNS within the cells [87]. Contradictory to chemiluminescence, florescent techniques have a higher accuracy, specificity, and reproducibility rate for intracellular ROS [88]. However, the utilization of florescent probes requires expensive equipment. The data generated does not quantify ROS but rather is indicative of the percentage of cells displaying a high ROS activity. An example of florescent probe used is 2,7-dichlorofluorescein diacetate (H2DCFDA) that penetrates the cells and indicates H_2O_2 concentrations, as H_2O_2 de-esterifies in the presence of DCFH and forms highly fluorescent 2,7-dichlorofluorescein (DCF) [89]. DCF fluoresces and this can be measured and indicates of formation of intracellular levels of hydrogen peroxide. Dihydroethidium/hydroethidine (DHE) is a non-florescent probe that is oxidized by a variety of reactive oxygen and nitrogen species. This probe is primarily used to visualize $O_2^{-\bullet}$ production [90].

5.3 Measurement of total antioxidants

Total antioxidant capacity (TAC) highlights the crucial role of antioxidant enzymes in counterbalancing ROS generation, and therefore can be a powerful tool in determining the redox status of a sample [91]. This measurement may give more relevant biological information compared to that obtained by the measurement of individual components (e.g., SOD, CAT, GPx) as it considers the cumulative effect of all antioxidants present in plasma and body fluids [92]. The total antioxidant assay relies on the ability of antioxidants in the sample to inhibit the oxidation of 2,20-azino-di-3-ethylbenzthiazoline sulphonate (ABTS) to ABTS+ by metmyoglobin. Briefly, metmyoglobin and hydrogen peroxide produced ferryl-myoglobin radical, which oxidized the ABTS (2, 2′-azino-bis (3-ethylbenzthiazoline-6-sulfonic acid) to produce a radical cation, ABTS+, a soluble chromogen that is green in color [93]. Then, the oxidized ABTS radical cation ($ABTS{\bullet}{+}$) reacts with antioxidants to reduce the ABTS radical and lose its bluish-green color [94]. This reaction may be monitored spectrophotometrically with a spectrophotometer, a common laboratory instrument. $ABTS{\bullet} +$ has several characteristics that make it suitable for colorimetric assays; it has several absorbance peaks at different wavelengths [94], it has a high extinction coefficient, its solubility in water is high, and it is Also soluble in organic media. This assay is often referred to as the Trolox equivalent antioxidant capacity (TEAC) assay. The reactivity of the various antioxidants tested is compared to that of Trolox, which is a water-soluble analog of vitamin E [95].

6. ROS in male reproductive diseases

Sperm require an endogenous level of ROS for normal sperm function and "optimum" concentrations are not yet fully determined; therefore, a better approach could be suggested, which could be to measure the downstream effects of OS on sperm, such as oxidative DNA damage or lipid peroxidation (LPO), since these have been shown to directly inhibit sperm function and fertilization [96]. Below, we outline the strengths and limitations of measuring of some assays used to measure the activity of these downstream markers.

6.1 ROS and LPO in sperm

LPO is one of the main consequences of OS and high concentrations of seminal ROS, this is because the plasma membranes of spermatozoa contain abundant polyunsaturated fatty acids (PUFA) [97]. LPO has a two-fold effect on fertility: it decreases motility, thereby reducing the number of sperm that reach the oocyte and decreases membrane fluidity necessary for sperm-oocyte fusion [98]. The lipid peroxidation cascades result in the formation and accumulation of lipid aldehydes, including malondialdehyde (MDA) and 4-hydroxy-2-nonenal (4HNE) that are capable of disrupting sperm function through the formation of adducts with key proteins and DNA [96].

6.1.1 Thiobarbiuric acid-malondialdehyde (TBA-MDA)

In particular, MDA, one of the by-products of lipid breakdown, reacts with TBA and can be used in biochemical assays as a marker to monitor the degree of peroxidative damage sustained by spermatozoa [99]. Although its clinical use has been questioned, the results of this test show an excellent correlation with the degree of impairment of sperm function in terms of motility and sperm-oocyte fusion ability, thus deeming the test reliable and accurate in assessing male infertility [96].

6.1.2 Hydroxy-2-nonenal (4HNE)

Another cytotoxic break down product of LPO is 4HNE, a highly reactive lipid aldehyde that can react with proteins and nucleic acids, as well as other lipids present in spermatozoa [100]. Increased concentrations of 4HNE in human sperm have been shown to induce impaired motility, compromised membrane integrity, and reduced oocyte-binding ability, with a significant negative impact on fertilization potential [101]. 4HNE can be quantified in semen samples using antibodies that detect and attach to the lipids, which are then quantified using ELISA assay or using either western blot and immunohistochemistry analysis [96]. Although quantifying 4HNE in sperm holds some promise in infertility diagnostics, the need for highly specialized and pricey equipment prevents its clinical application. Therefore, further attempts should be made to develop stable, highly specific probes for the detection of lipid peroxidation in sperm. In this regard, probe BODIPY-C11 has recently been developed for determining peroxidative damage in human sperm. Elevated fluorescence by BODYIP-C11 has been correlated with increased ROS generation and genital urinary tract infection and has been negatively associated with sperm motility [102]. However, BODPIY-C11 may overrepresenting lipid peroxidation of cells [96].

6.2 ROS: induced sperm DNA damage

In general, DNA damage has been shown to decrease semen parameters (motility, viability, etc.) and reduce the success of *in vitro* fertilization (IVF). The main tests used for the assessment of DNA damage in spermatozoa induced by oxidative stress are summarized in **Table 1**; however, 8-hydroxy-2′-deoxyguanosine (8OHdG) deserves a special mention. 8-hydroxy-2′-deoxyguanosine (8OHdG) is a DNA base adduct, commonly used as a biomarker of oxidative stress due to its association with nuclear and mitochondrial DNA damage [103]. Formation of 8OHdG in semen samples has been correlated with increases in DNA fragmentation, chromatin retention, decreases in sperm motility, and fertilization rates, thus making it a promising potential biomarker for oxidative stress damage and male infertility [104]. Currently, 8OHdG can be quantified in semen samples using light microscopy, fluorescence microscopy and flow cytometry [105]. The main limitations are due to long and difficult protocols, as well as the need for large and expensive equipment.

6.3 ROS and TAC of semen

TAC in the seminal plasma can be measured using an enhanced chemiluminescence assay [91]. The fact that neither ROS alone nor TAC alone can adequately quantify seminal OS led to the logical conclusion that combining these two variables

Technique.	Assay principle	Detection method
TUNEL assay	DNA fragmentation, adds labeled nucleotides to free DNA ends, and single- and double-strand DNA breaks	Flow cytometry/fluorescence microscopy
Comet assay	Intact DNA stays in the head, DNA fragments form tails, evaluates DNA integrity, and single- and double-strand DNA breaks	fluorescence microscopy
Sperm chromatin structure assay	Acid DNA denaturation, single- or double-strand breaks	Flow cytometry
DNA breakage Detection-fluorescence in situ hybridization	DNA breaks and denatures nicked DNA	fluorescence microscopy and image analyzer

Table 1.
Principal assays for the assessment of ROS-induced DNA damage in spermatozoa.

may be a better index for the diagnosis of the overall OS affecting spermatozoa. The ROS-TAC score is an accurate measure of seminal OS, and low ROS-TAC scores indicate high seminal OS [91].

6.4 Current clinical measures of ROS in semen

The advanced examinations section of the most recent sixth edition of the WHO manual for the assessment of human semen has been updated to include two of the three currently known clinically available measurements of ROS in semen, Luminol and MiOXYSIS [96]. Currently, these measures are only performed in men if abnormalities are observed during standard semen analysis, or when couples undergoing ART have experienced reduced fertilization and embryo development rates, or repeated implantation failure [106].

6.4.1 Luminol chemiluminescence

Chemiluminescent assays have been used to show a negative association between an increase in ROS levels and sperm parameters. These parameters include sperm motility, viability, morphology, and concentration [107]. A variety of factors may influence the data generated by the chemiluminescence assays, which include the presence of leukocytes in the sperm sample, sperm incubation time, the pH, and contamination of the seminal plasma [108]. It should also be noted that both sample and probe concentrations also affect luminescence; thus, it is important to have a fixed probe concentration for varying concentrations of sperm. However, chemiluminescence assays are sensitive, which is ideal as sperm generally produce low concentrations of ROS [109].

6.4.2 MiOXSYS system™

The MiOXSYS™ is a highly specific *in vitro* diagnostic tool used to measure static oxidation-reduction potential (sORP) in human semen [110]. The system works by measuring the transfer of electrons between oxidants and reductants within fresh semen samples to ultimately calculate total oxidant and antioxidant activity present within the ejaculate. Samples of high sORP levels indicate an imbalance that suggests the presence of OS [110]. Unlike chemiluminescence, where ROS levels have a significantly short half-life, ORP measurements are stable for up to 120 min. The assay requires only 30 μL of fresh or frozen semen sample and produces results in less than 5 min, making it a popular choice in both clinical and research settings [83]. Additionally, MiOXSYS sORP's ability to measure all oxidants and reductants makes it clinically meaningful in the diagnosis of cases of male infertility that is associated with high ROS levels [111].

6.4.3 OxiSperm®

OxiSperm® measures the presence of excess O_2^- in sperm, seminal plasma, and whole semen. The assay is largely based on nitroblue tetrazolium (NTB) staining in which the yellow NBT molecule is reduced into the insoluble blue crystals, called formazan, in the presence of O_2. This assay uses a gel that shows varying intensities once reacted that can be categorized as low, medium, or high as per the color scale [112]. However, the semi-qualitative measurement of OxiSperm® creates considerable opportunity for introducing bias into results while also showing low assay precision, as differences in color perception between scientists could result in different interpretations [96]. This may partially explain the absence of OxiSperm kits from inclusion in the latest sixth edition WHO guidelines under the recommended assessments of ROS.

7. ROS in female reproductive diseases

OS also affects the body fluids, tissues, and organs of the female reproductive system; not only, a host of female infertility diseases are attributed to the production of pathological ROS levels [113]. Concentrations of ROS may also play a major role both in the implantation and fertilization of eggs [114]. A number of OS biomarkers have been investigated, including SOD, LPO, oxidative DNA adducts, and TAC, both in peritoneal fluid and in whole blood and in the placenta. The principal biomarkers are summarized in **Table 2**.

7.1 ROS in ovaries and follicular fluid

Various studies have confirmed the role of ROS in follicular maturation, folliculogenesis, function of the corpus luteum, as well as ovulation [22]. The environment of the follicular fluid is thought to play a critical role in oocyte maturation and the eventual development of an embryo [16]. The main sources of ROS in the follicular fluid microenvironment are macrophages, leukocytes, and cytokines [115]. At the same time, a crosstalk, mediated by ROS, cytokines, and vascular endothelial growth factor (VEGF), important for ovarian folliculogenesis and embryo formation has been demonstrated [16]. Accordingly, ROS are involved in follicular growth in part by regulating angiogenesis. Increased ROS levels have been associated with poor oocyte quality, low fertilization rate, and impaired embryo development [115]. Furthermore, any imbalance between the cytokines and angiogenesis factors could result in implantation failure and pregnancy loss [116]. Normal ovaries express many of the common biomarkers of OS [117]. Markers of OS such as Cu-SOD, Zn-SOD, Mn-SOD, GPx, γ glutamyl synthetase, and lipid peroxides have been measured by immunohistochemical staining, mRNA expression, and thiobarbituric acid methods [118]. Additionally, 8-OHDG has become an important marker for OS in ovaries, as causes base mutation and mismatches in DNA replication [16].

Biomarkers	Methodology
SOD, CAT, GPx	Reverse transcription-polymerase chain reaction
TAC	Chemiluminescence assay
LPO, MDA, coniugated dienes	Thiobarbituric acid method
Oxidative DNA adducts	8-hydroxy-2′-deoxyguanosine measurement
Plasma and red blood cell, glutathione content	Colorimetric assay, fluorometric assay
Superoxide anion, hydrogen peroxide peroxynitrite	Spectrophotometry/flow cytometry

Table 2.
Principal biomarkers of ROS in the female reproductive system.

7.2 ROS in amniotic/placental fluid

During pregnancy, the mother and fetus can be exposed to high levels of OS due to the higher metabolic demand of the growing fetus [16]. Superoxide anions produced by the placental mitochondria appear to be a major source of ROS and LPO contributing to OS in the placenta [119]. However, the placenta gradually adapts to this environment and returns to normal under the action of antioxidant activity [120]. These studies showed that limited levels of ROS are necessary to maintain physiological function, but when present in higher concentrations, ROS can have deleterious effects. The placenta experiences a heightened level of OS in certain pathologic conditions of pregnancies, including gestational diabetes, fetal growth restriction, preeclampsia, and miscarriage [119]. The biomarkers of OS in the placenta and amniotic fluid include LHP, intracellular ROS, TAC, and DNA adducts-8-OHDG [71]. Another biomarker of ROS is LPO, which is the oxidative destruction of polyunsaturated fatty acids in the plasma membrane [16]. This leads to increased membrane permeability, degraded membrane integrity, structural damage of the DNA, and cell death.

7.3 ROS in peritoneal fluid

Peritoneal fluid bathes the pelvic cavity, uterus, fallopian tubes, and ovaries. It may be an important factor controlling the peritoneal microenvironment influencing the development of some pathologies such as preeclampsia ad endometriosis [121]. The peritoneal fluid of patients with endometriosis has been found to contain high concentrations of MDA, pro-inflammatory cytokines (IL-6, TNF-α, and IL-1β), angiogenic factors (IL-8 and VEGF), monocyte chemoattractant protein-1 (MCP-1), and oxidized LDL (ox-LDL) [122]. Pro-inflammatory and chemotactic cytokines play central roles in the recruitment and activation of phagocytic cells, which are the main producers of ROS and RNS [73]. ROS also appears to play a critical role in the endothelial dysfunction associated with preeclampsia. Increased ROS concentrations in patients with preeclampsia have been proved by the increased levels of MDA, an index of lipid peroxidation [16]. Additionally, levels of TNF-α and oxLDL are increased in preeclampsia and have been shown to activate the endothelial isoform of NAD(P)H oxidase, ultimately resulting in increased levels of the SO anion [123]. Affected women also have decreased TAC and placental GPx [16].

8. Conclusion

In conclusion with this chapter, we hope to have given a broad description of the scientific bibliography on the role played by ROS in physiological and pathological conditions in both men and women. New studies aimed at understanding whether it is possible to reduce oxidative stress with nutraceuticals, physical exercise, or diet, for example, in order to delay and/or limit the onset of pathologies of the reproductive system, are obviously indispensable. Some studies for example have evaluated the effects of diet in mice model endometriosis, testicular damage cyclofasphamide-induced, testicular growth, urethral syndrome, chronic prostatitis/chronic pelvic pain syndrome, and more [124–128]. Further, studies will also be essential to discover new and advanced methods of diagnosis through biomarkers in order to make the diagnosis ever earlier and more precise.

Acknowledgements

These authors want to thank to Dr. Livia Interdonato and Alessia Arangia for their incredible support.

Conflict of interest

The authors declare no conflict of interest.

Author details

Ramona D'Amico[1], Salvatore Cuzzocrea[1], Marika Cordaro[2*] and Rosanna Di Paola[3]

1 Department of Chemical, Biological, Pharmaceutical and Environmental Sciences, University of Messina, Messina, Italy

2 Department of Biomedical and Dental Sciences and Morphofunctional Imaging, University of Messina, Messina, Italy

3 Department of Veterinary Sciences, University of Messina, Messina, Italy

*Address all correspondence to: cordarom@unime.it

References

[1] Ozkocer SE, Konac E. The current perspective on genetic and epigenetic factors in sperm maturation in the epididymis. Andrologia. 2021;**53**(3):e13989

[2] Das A, Roychoudhury S. Reactive Oxygen Species in the Reproductive System: Sources and Physiological Roles. In: Kesari KK, Roychoudhury S, editors. Oxidative Stress and Toxicity in Reproductive Biology and Medicine. Advances in Experimental Medicine and Biology. Vol. 1358. Cham: Springer; 2022. DOI: 10.1007/978-3-030-89340-8_2

[3] Beygi Z et al. Role of oxidative stress and antioxidant supplementation in male fertility. Current Molecular Medicine. 2021;**21**(4):265-282

[4] Kongmanas K et al. Accumulation of seminolipid in sertoli cells is associated with increased levels of reactive oxygen species and male subfertility: Studies in aging arsa null male mice. Antioxidants. 2021;**10**(6):912

[5] Dutta S, Henkel R, Sengupta P, Agarwal A, et al. Physiological role of ROS in sperm function. Male infertility: Contemporary clinical approaches, andrology. ART and Antioxidants. 2020:337-345

[6] Naseer Z, Raza S, Ahmad E, Aksoy M. Use of antioxidants to improve the reproductive performance in model laboratory animals. The Role of Exogenous Antioxidants in Enhancing Reproductive Function and Performance, Chapter 1. 2022

[7] Baskaran S et al. Reactive oxygen species in male reproduction: A boon or a bane? Andrologia. 2021;**53**(1):e13577

[8] Sullivan R, Mieusset R. The human epididymis: Its function in sperm maturation. Human Reproduction Update. 2016;**22**(5):574-587

[9] Peña FJ et al. Redox regulation and oxidative stress: The particular case of the stallion spermatozoa. Antioxidants. 2019;**8**(11):567

[10] Aitken RJ. Reactive oxygen species as mediators of sperm capacitation and pathological damage. Molecular Reproduction and Development. 2017;**84**(10):1039-1052

[11] Takei GL et al. Activation of cAMP-dependent phosphorylation pathways is independent of ROS production during mouse sperm capacitation. Molecular Reproduction and Development. 2021;**88**(8):544-557

[12] GONZÁLEZ S, Ruturi S, et al. Cortisol modulates Ca^{2+} signaling and acrosome reaction in human sperm. Andrology. 2023;**11**(1):134-142

[13] Balestrini PA et al. Membrane hyperpolarization abolishes calcium oscillations that prevent induced acrosomal exocytosis in human sperm. The FASEB Journal. 2021;**35**(6):e21478

[14] Aitken RJ. Impact of oxidative stress on male and female germ cells: Implications for fertility. Reproduction. 2020;**159**(4):R189-R201

[15] Engel KM et al. Metabolomic profiling reveals correlations between spermiogram parameters and the metabolites present in human spermatozoa and seminal plasma. PLoS One. 2019;**14**(2):e0211679

[16] Lu J et al. A novel and compact review on the role of oxidative stress in female reproduction. Reproductive Biology and Endocrinology. 2018;**16**:80

[17] Alam F et al. Association of oxidative stress with female infertility-a case control study. JPMA. The Journal of the Pakistan Medical Association. 2019;**69**(5):627

[18] Ávila J et al. Oxidative stress in granulosa-lutein cells from in vitro fertilization patients. Reproductive Sciences. 2016;**23**:1656-1661

[19] Ahmed TA et al. Oocyte aging: The role of cellular and environmental factors and impact on female fertility. Cell Biology and Translational Medicine. Stem Cells in Regenerative Medicine. 2020;**8**:109-123

[20] Campos-Rentería AC et al. Evaluation of antioxidant enzymes in the female genital fluid of the bat Corynorhinus mexicanus during sperm storage. Animal Biology. 2023;**1**(aop):1-14

[21] Goutami L, Jena SR, Swain A, Samanta L, et al. Pathological role of reactive oxygen species on female reproduction. In: Oxidative Stress and Toxicity in Reproductive Biology and Medicine: A Comprehensive Update on Male Infertility Volume II. Cham: Springer International Publishing; 2022. pp. 201-220

[22] Wang L et al. Oxidative stress in oocyte aging and female reproduction. Journal of Cellular Physiology. 2021;**236**(12):7966-7983

[23] Liang J et al. Reactive oxygen species and ovarian diseases: Antioxidant strategies. Redox Biology. 2023:102659

[24] Podolak A, Woclawek-Potocka I, Lukaszuk K. The role of mitochondria in human fertility and early embryo development: What can we learn for clinical application of assessing and improving mitochondrial DNA? Cell. 2022;**11**(5):797

[25] Shrestha HK. Loss of luteal sensitivity to luteinizing hormone underlies luteolysis in cattle: A hypothesis. Reproductive Biology. 2021;**21**(4):100570

[26] Mahmoud AA, Elfiky AM, Abo-Zeid FS. The anti-androgenic effect of quercetin on hyperandrogenism and ovarian dysfunction induced in a dehydroepiandrosterone rat model of polycystic ovary syndrome. Steroids. 2022;**177**:108936

[27] Przygrodzka E, Plewes MR, Davis JS. Luteinizing hormone regulation of inter-organelle communication and fate of the corpus luteum. International Journal of Molecular Sciences. 2021;**22**(18):9972

[28] Kawaguchi S, Sakumoto R, Okuda K. Induction of the expressions of antioxidant enzymes by luteinizing hormone in the bovine corpus luteum. The Journal of Reproduction and Development. 2013;**59**(3):219-224

[29] World Health Organization. WHO Laboratory Manual for the Examination and Processing of Human Semen. Geneva: World Health Organization; 2010

[30] Agarwal A, Tvrda E, Sharma R. Relationship amongst teratozoospermia, seminal oxidative stress and male infertility. Reproductive Biology and Endocrinology. 2014;**12**:45

[31] Du Plessis SS et al. Contemporary evidence on the physiological role of reactive oxygen species in human sperm function. Journal of Assisted Reproduction and Genetics. 2015;**32**(4):509-520

[32] Agarwal A, Prabakaran S, Allamaneni SS. Relationship between oxidative stress, varicocele and infertility: A meta-analysis. Reproductive Biomedicine Online. 2006;**12**(5):630-633

[33] Makker K, Agarwal A, Sharma R. Oxidative stress & male infertility. The Indian Journal of Medical Research. 2009;**129**(4):357-367

[34] Cho CL, Esteves SC, Agarwal A. Novel insights into the pathophysiology of varicocele and its association with reactive oxygen species and sperm DNA fragmentation. Asian Journal of Andrology. 2016;**18**(2):186-193

[35] Agarwal A et al. Effect of cell phone usage on semen analysis in men attending infertility clinic: An observational study. Fertility and Sterility. 2008;**89**(1):124-128

[36] Alshahrani S, McGill J, Agarwal A. Prostatitis and male infertility. Journal of Reproductive Immunology. 2013;**100**(1):30-36

[37] Irez T et al. Cytokines and adipokines in the regulation of spermatogenesis and semen quality. Chemical Biology Letters. 2020;**7**:131-140

[38] Sanocka D, Kurpisz M. Reactive oxygen species and sperm cells. Reproductive Biology and Endocrinology. 2004;**2**(1):12

[39] Tremellen, K., Oxidative stress and male infertility—A clinical perspective. Human Reproduction Update, 2008. 14(3): p. 243-258.

[40] Okonofua FE et al. Causes and risk factors for male infertility: A scoping review of published studies. International Journal of General Medicine. 2022;**15**:5985-5997

[41] Liu G et al. Restraint stress delays endometrial adaptive remodeling during mouse embryo implantation. Stress. 2015;**18**(6):699-709

[42] Lipsky BA, Byren I, Hoey CT. Treatment of bacterial prostatitis. Clinical Infectious Diseases. 2010;**50**(12):1641-1652

[43] Krupp K, Madhivanan P. Antibiotic resistance in prevalent bacterial and protozoan sexually transmitted infections. Indian Journal of Sexually Transmitted Diseases and AIDS. 2015;**36**(1):3-8

[44] Magri V et al. Efficacy of repeated cycles of combination therapy for the eradication of infecting organisms in chronic bacterial prostatitis. International Journal of Antimicrobial Agents. 2007;**29**(5):549-556

[45] Oliphant CM, Green GM. Quinolones: A comprehensive review. American Family Physician. 2002;**65**(3):455-464

[46] Costello LC et al. Role of zinc in the pathogenesis and treatment of prostate cancer: Critical issues to resolve. Prostate Cancer and Prostatic Diseases. 2004;**7**(2):111-117

[47] Krause W, Bohring C. Male infertility and genital chlamydial infection: Victim or perpetrator? Andrologia. 2003;**35**(4):209-216

[48] Workowski KA, Berman S. Sexually transmitted diseases treatment guidelines, 2010. Morbidity and Mortality Weekly Report. 2010;**59**(Rr-12):1-110

[49] Dutta S et al. Staphylococcal infections and infertility: Mechanisms and management. Molecular and Cellular Biochemistry. 2020;**474**(1-2):57-72

[50] Agarwal A et al. Male oxidative stress infertility (MOSI): Proposed terminology and clinical practice guidelines for Management of Idiopathic Male Infertility. World Journal of Men's Health. 2019;**37**(3):296-312

[51] Koppers AJ, Garg ML, Aitken RJ. Stimulation of mitochondrial reactive oxygen species production by unesterified, unsaturated fatty acids in defective human spermatozoa. Free Radical Biology & Medicine. 2010;**48**(1):112-119

[52] Iwasaki A, Gagnon C. Formation of reactive oxygen species in spermatozoa of infertile patients. Fertility and Sterility. 1992;**57**(2):409-416

[53] Panner Selvam MK, Sengupta P, Agarwal A. In: Arafa M et al., editors. Sperm DNA Fragmentation and Male Infertility, in Genetics of Male Infertility: A Case-Based Guide for Clinicians. Cham: Springer International Publishing; 2020. pp. 155-172

[54] Dutta S et al. Physiological role of ROS in sperm function. In: Parekattil SJ, Esteves SC, Agarwal A, editors. Male Infertility: Contemporary Clinical Approaches, Andrology, ART and Antioxidants. Cham: Springer International Publishing; 2020. pp. 337-345

[55] Sengupta P, Dutta S. Hormones in male reproduction and fertility. Asian Pacific Journal of Reproduction. 2019;**8**(5):187-188

[56] Dutta S et al. Oxidative stress, testicular inflammatory pathways, and male reproduction. International Journal of Molecular Sciences. 2021;**22**(18)

[57] Wells D et al. Expression of genes regulating chromosome segregation, the cell cycle and apoptosis during human preimplantation development. Human Reproduction. 2005;**20**(5):1339-1348

[58] Sakkas D, Alvarez JG. Sperm DNA fragmentation: Mechanisms of origin, impact on reproductive outcome, and analysis. Fertility and Sterility. 2010;**93**(4):1027-1036

[59] Tesarik J, Greco E, Mendoza C. Late, but not early, paternal effect on human embryo development is related to sperm DNA fragmentation. Human Reproduction. 2004;**19**(3):611-615

[60] Tesarík J et al. Activation of nucleolar and extranucleolar RNA synthesis and changes in the ribosomal content of human embryos developing in vitro. Journal of Reproduction and Fertility. 1986;**78**(2):463-470

[61] Kuroda S et al. Early and late paternal effects of reactive oxygen species in semen on embryo development after intracytoplasmic sperm injection. Systems Biology in Reproductive Medicine. 2020;**66**(2):122-128

[62] Zhao J et al. Whether sperm deoxyribonucleic acid fragmentation has an effect on pregnancy and miscarriage after in vitro fertilization/intracytoplasmic sperm injection: A systematic review and meta-analysis. Fertility and Sterility. 2014;**102**(4):998-1005.e8

[63] Trevelyan SJ et al. Structure-based mechanism of preferential complex formation by apoptosis signal-regulating kinases. Science Signaling. 2020;**13**(622):eaay6318

[64] Shukla KK, Mahdi AA, Rajender S. Apoptosis, spermatogenesis and male infertility. Frontiers in Bioscience (Elite Edition). 2012;**4**(2):746-754

[65] Calivarathan L et al. Environmental Contaminants and Male Reproductive Health: Past, Present and Future. In: Environmental Studies and Climate Change. CRC Press; 2022. p. 361-372

[66] Gupta S, Agarwal A, Banerjee J, Alvarez JG. The role of oxidative stress in spontaneous abortion and recurrent pregnancy loss: A systematic review. Obstetrical & Gynecological Survey. 2007;**62**(5):335-347

[67] Ota H, Igarashi S, Tanaka T. Xanthine oxidase in eutopic and ectopic endometrium in endometriosis and adenomyosis. Fertility and Sterility. 2001;**75**(4):785-790

[68] Gupta S et al. Role of oxidative stress in endometriosis. Reproductive Biomedicine Online. 2006;**13**(1):126-134

[69] Kajihara H et al. New insights into the pathophysiology of endometriosis: From chronic inflammation to danger signal. Gynecological Endocrinology. 2011;**27**(2):73-79

[70] McCubrey JA, Lahair MM, Franklin RA. Reactive oxygen species-induced activation of the MAP kinase signaling pathways. Antioxidants & Redox Signaling. 2006;**8**(9-10):1775-1789

[71] Agarwal A, Gupta S, Sikka S. The role of free radicals and antioxidants in reproduction. Current Opinion in Obstetrics & Gynecology. 2006;**18**(3):325-332

[72] Madazli R et al. The plasma and placental levels of malondialdehyde, glutathione and superoxide dismutase in pre-eclampsia. Journal of Obstetrics and Gynaecology. 2002;**22**(5):477-480

[73] Matsubara K et al. Nitric oxide and reactive oxygen species in the pathogenesis of preeclampsia. International Journal of Molecular Sciences. 2015;**16**(3):4600-4614

[74] Hubel CA et al. Lipid peroxidation in pregnancy: New perspectives on preeclampsia. American Journal of Obstetrics and Gynecology. 1989;**161**(4):1025-1034

[75] Uzun H et al. Circulating oxidized low-density lipoprotein and paraoxonase activity in preeclampsia. Gynecologic and Obstetric Investigation. 2005;**60**(4):195-200

[76] Behrman HR et al. Oxidative stress and the ovary. Journal of the Society for Gynecologic Investigation. 2001;**8**(1 Suppl Proceedings):S40-S42

[77] Raijmakers MTM et al. NAD(P)H oxidase associated superoxide production in human placenta from normotensive and pre-eclamptic women. Placenta. 2004;**25**:S85-S89

[78] Walsh SW. Eicosanoids in preeclampsia. Prostaglandins, Leukotrienes, and Essential Fatty Acids. 2004;**70**(2):223-232

[79] Klemmensen A et al. Intake of vitamin C and E in pregnancy and risk of pre-eclampsia: Prospective study among 57 346 women. BJOG. 2009;**116**(7):964-974

[80] Liu G et al. Restraint stress alters immune parameters and induces oxidative stress in the mouse uterus during embryo implantation. Stress. 2014;**17**(6):494-503

[81] Xiaoyan S et al. Follicle stimulating hormone inhibits the expression of p53 up-regulated modulator of apoptosis induced by reactive oxygen species through PI3K/AKT in mouse granulosa cells. Physiological Research. 2020;**69**(4):687

[82] Denefil OV et al. The peculiarities of morphological changes of rats'ovary and biochemical state under the damage with different doses of lead acetate. Wiadomości Lekarskie. 2022;**75**(2):377-382

[83] Ayad B et al. Oxidative stress and male infertility: Evidence from a research perspective. Human Reproductive Health. 2022;**4**:822257

[84] Agarwal A, Qiu E, Sharma R. Laboratory assessment of oxidative

stress in semen. Arab Journal of Urology. 2018;**16**(1):77-86

[85] Gosalvez J, Tvrda E, Agarwal A. Free radical and superoxide reactivity detection in semen quality assessment: Past, present, and future. Journal of Assisted Reproduction and Genetics. 2017;**34**(6):697-707

[86] Agarwal A, Allamaneni SS, Said TM. Chemiluminescence technique for measuring reactive oxygen species. Reproductive Biomedicine Online. 2004;**9**(4):466-468

[87] Zhang X et al. ROS/RNS and base dual activatable merocyanine-based NIR-II fluorescent molecular probe for in vivo biosensing. Angewandte Chemie (International Ed. in English). 2021;**60**(50):26337-26341

[88] Hans C et al. 2′,7′-Dichlorofluorescein (DCF) or 2′,7′-dichlorodihydrofluorescein diacetate (DCFH2-DA) to measure reactive oxygen species in erythrocytes. Biomedicine & Pharmacotherapy. 2021;**138**:111512

[89] Yao H, Jockusch RA. Fluorescence and electronic action spectroscopy of mass-selected gas-phase fluorescein, 2′,7′-dichlorofluorescein, and 2′,7′-difluorofluorescein ions. The Journal of Physical Chemistry. A. 2013;**117**(6):1351-1359

[90] Kalyanaraman B et al. HPLC-based monitoring of products formed from hydroethidine-based fluorogenic probes--the ultimate approach for intra- and extracellular superoxide detection. Biochimica et Biophysica Acta. 2014;**1840**(2):739-744

[91] Gupta S et al. Total antioxidant capacity-relevance, methods and clinical implications. Andrologia. 2021;**53**(2):e13624

[92] Apak R et al. Antioxidant activity/capacity measurement. 3. Reactive oxygen and nitrogen species (ROS/RNS) scavenging assays, oxidative stress biomarkers, and chromatographic/chemometric assays. Journal of Agricultural and Food Chemistry. 2016;**64**(5):1046-1070

[93] Hosen MB et al. Oxidative stress induced sperm DNA damage, a possible reason for male infertility. International Journal of Reproductive BioMedicine. 2015;**13**(9):525-532

[94] Cano A et al. ABTS/TAC methodology: Main milestones and recent applications. PRO. 2023;**11**(1):185

[95] Barclay LRC, Locke SJ, MacNeil JM. Autoxidation in micelles. Synergism of vitamin C with lipid-soluble vitamin E and water-soluble Trolox. Canadian Journal of Chemistry. 1985;**63**(2):366-374

[96] Castleton PE, Deluao JC, Sharkey DJ, McPherson NO. Measuring reactive oxygen species in semen for male preconception care: A scientist perspective. Antioxidants (Basel). 2022;**11**(2):264

[97] Agarwal A et al. Sperm vitality and necrozoospermia: Diagnosis, management, and results of a global survey of clinical practice. World Journal of Men's Health. 2022;**40**(2):228-242

[98] Walters JLH et al. Mechanistic insight into the regulation of lipoxygenase-driven lipid peroxidation events in human spermatozoa and their impact on male fertility. Antioxidants (Basel). 2020;**10**(1):43

[99] du Plessis SS et al. Oxidative phosphorylation versus glycolysis: What fuel do spermatozoa use? Asian Journal of Andrology. 2015;**17**(2):230-235

[100] Walters JLH et al. Oxidative stress in the male germline: A review of novel strategies to reduce 4-Hydroxynonenal production. Antioxidants (Basel). 2018;7(10):132

[101] Nowicka-Bauer K, Nixon B. Molecular changes induced by oxidative stress that impair human sperm motility. Antioxidants (Basel). 2020;**9**(2):134

[102] Ghantabpour T et al. The effect of astaxanthin on motility, viability, reactive oxygen species, apoptosis, and lipid peroxidation of human spermatozoa during the freezing-thawing process. Biopreservation and Biobanking. 2022;**20**(4):367-373

[103] Rahman MF et al. Effects of elevated temperature on 8-OHdG expression in the American oyster (*Crassostrea virginica*): Induction of oxidative stress biomarkers, cellular apoptosis, DNA damage and gammaH2AX signaling pathways. Fish and Shellfish Immunology Reports. 2023;**4**:100079

[104] De Iuliis GN et al. DNA damage in human spermatozoa is highly correlated with the efficiency of chromatin remodeling and the formation of 8-hydroxy-2′-deoxyguanosine, a marker of oxidative stress. Biology of Reproduction. 2009;**81**(3):517-524

[105] Aitken RJ, Bakos HW. Should we be measuring DNA damage in human spermatozoa? New light on an old question. Human Reproduction. 2021;**36**(5):1175-1185

[106] Korneyev IA. Oxidative stress and male infertility a clinical perspective. Urologiia. 2022;**1**:102-108

[107] Mannucci A et al. The impact of oxidative stress in male infertility. Frontiers in Molecular Biosciences. 2021;**8**:799294

[108] Dutta S, Majzoub A, Agarwal A. Oxidative stress and sperm function: A systematic review on evaluation and management. Arab Journal of Urology. 2019;**17**(2):87-97

[109] Agarwal A, Ahmad G, Sharma R. Reference values of reactive oxygen species in seminal ejaculates using chemiluminescence assay. Journal of Assisted Reproduction and Genetics. 2015;**32**(12):1721-1729

[110] Homa ST et al. A comparison between two assays for measuring seminal oxidative stress and their relationship with sperm DNA fragmentation and semen parameters. Genes (Basel). 2019;**10**(3):236

[111] Liu K et al. Correlation analysis of sperm DNA fragmentation index with semen parameters and the effect of sperm DFI on outcomes of ART. Scientific Reports. 2023;**13**(1):2717

[112] Castleton P et al. MiOXSYS® and OxiSperm® II assays appear to provide no clinical utility for determining oxidative stress in human sperm-results from repeated semen collections. Andrology. 2022

[113] Agarwal A, Gupta S, Sharma RK. Role of oxidative stress in female reproduction. Reproductive Biology and Endocrinology. 2005;**3**:28

[114] Agarwal A et al. The effects of oxidative stress on female reproduction: A review. Reproductive Biology and Endocrinology. 2012;**10**:49

[115] Valsamakis G, Chrousos G, Mastorakos G. Stress, female reproduction and pregnancy. Psychoneuroendocrinology. 2019;**100**:48-57

[116] Rizov M, Andreeva P, Dimova I. Molecular regulation and

role of angiogenesis in reproduction. Taiwanese Journal of Obstetrics & Gynecology. 2017;**56**(2):127-132

[117] Steller JG, Alberts JR, Ronca AE. Oxidative stress as cause, consequence, or biomarker of altered female reproduction and development in the space environment. International Journal of Molecular Sciences. 2018;**19**(12):3729

[118] Victoria M et al. Anti Mullerian hormone: More than a biomarker of female reproductive function. Journal of Gynecology Obstetrics and Human Reproduction. 2019;**48**(1):19-24

[119] Ruano CSM et al. The impact of oxidative stress of environmental origin on the onset of placental diseases. Antioxidants (Basel). 2022;**11**(1):106

[120] Bisson C et al. Preeclampsia pathophysiology and adverse outcomes during pregnancy and postpartum. Frontiers in Medicine (Lausanne). 2023;**10**:1144170

[121] Guo M et al. Mass cytometry analysis reveals a distinct immune environment in peritoneal fluid in endometriosis: A characterisation study. BMC Medicine. 2020;**18**(1):3

[122] Sipak-Szmigiel O et al. Serum and peritoneal fluid concentrations of soluble human leukocyte antigen, tumor necrosis factor alpha and interleukin 10 in patients with selected ovarian pathologies. Journal of Ovarian Research. 2017;**10**(1):25

[123] Negi R et al. Biochemical markers in the pathogenesis of preeclampsia: Novel link between placental growth factor and interleukin-6. Molecular and Cellular Biochemistry. 2022;**477**(6):1765-1774

[124] Krisiloff M. A dietary cure for prostatitis and the urethral syndrome. Infectious Diseases in Clinical Practice. 2002;**11**(3):107-110

[125] Martin GB et al. Effects of dietary zinc deficiency on the reproductive system of young male sheep: Testicular growth and the secretion of inhibin and testosterone. Journal of Reproduction and Fertility. 1994;**101**(1):87-96

[126] Hansen SO, Knudsen UB. Endometriosis, dysmenorrhoea and diet. European Journal of Obstetrics, Gynecology, and Reproductive Biology. 2013;**169**(2):162-171

[127] Parazzini F et al. Diet and endometriosis risk: A literature review. Reproductive Biomedicine Online. 2013;**26**(4):323-336

[128] Herati AS, Moldwin RM. Alternative therapies in the management of chronic prostatitis/chronic pelvic pain syndrome. World Journal of Urology. 2013;**31**(4):761-766

Chapter 7

Oxidative Stress and Male Fertility: Promising Role of Nutraceuticals

Zahid Naseer, Mudussar Nawaz, Ejaz Ahmad and Zia ur Rehman

Abstract

Oxidative stress is a key detrimental factor in male infertility under pathological or physiological conditions. A balanced oxidation-reduction process regulates the various functions linked to male fertility; however, oxidative stress leads to temporary infertility by affecting the hormonal pattern, sexual behavior, testicular milieu, functioning of accessory sex glands, and sperm quality. Currently, nutraceuticals are a common and popular way to mitigate the male fertility issues of pre-testicular, testicular, and post-testicular etiologies. Nutraceuticals possess multi-nutritional factors that improve metabolic activity, regulating hormonal profile, and sperm production. In addition, the antioxidant property of nutraceuticals agents combats oxidative stress, thus improving the hormonal release pattern, sexual behavior, testicular environment, and sperm quality in males.

Keywords: male infertility, oxidative stress, nutraceuticals, sperm quality, reproduction

1. Introduction

The incidences of male infertility have been rapidly increased during the last five decades perhaps due to the industrialization and mechanization. The rising surge in male infertility is multifactorial and could be linked to inevitable conditions such as congenital assaults (cryptorchidism), infections, age, varicocele, etc. or environmental contaminants, obesity, use of recreational drugs, occupational hazards, and irrational use of antimicrobial drugs [1]. The extent of infertility varies from region to region depending on the extensive industrialization, high mechanization, frequency of exposure to chemicals or radiation, and less physical activity [2]. Infertility is observed at pre-testicular, testicular, or post-testicular levels due to the mechanism of action of each influencing factor. These physiological or pathological abnormalities result in alteration in the activity of the hypothalamus, pituitary, or gonads is probably linked to the production of free radicals, which are manifested in the form of low libido, erectile dysfunction, debilitated accessory sex glands, impaired spermatogenesis, and sperm functions [3]. The onset of these conditions is chronic rather than acute states, which originated maximally due to oxidative stress. Oxidative stress is linked to physiological and pathological conditions, and its prevention or treatment is also long-term concerning its occurrence [4].

The nutrition is essential to body health since a healthy and balanced diet prevents the occurrence or cure different disease conditions. The use of nutraceuticals as potential ailments for male infertility has been recognized in the old world for centuries [5]. Every passing day, a new compound is being included in the medicinal list with its bioactive ingredients or antioxidant potential. With the understanding of oxidative damage to sexual behavior, reduced spermatogenesis, and low sperm quality, numerous nutraceuticals have been focused on their antioxidants potential for enhancing male fertility [6, 7]. However, utilizing a single agent with multiple properties could be better option for infertile males. On the other hand, patients are unaware about the potential effect of any therapeutic agent on the hypothalamus, pituitary, testes, or extra-testicular parts. This chapter proposed an overview of the underlying mechanism of pros and cons of oxidative stress on the male reproductive system and countering effects of nutraceuticals with antioxidant potential against pre-testicular, testicular, and post-testicular etiologies.

1.1 What is male fertility? Definitions

Male reproductive health is important for the maintenance of sexual activities including libido, optimum sperm production, copulation, and fertilizing ability during sexual life. All these processes are necessary for maintaining the possible genetic potential and progeny. Previously, the male reproductive health has been defined in different ways.

Generally, the male fertility in humans is defined as the ability of a person to impregnate a woman and produce offspring.

- The American Society for Reproductive Medicine defines male fertility as "the capacity of a man to produce normal sperm in sufficient quantities and with appropriate motility and morphology to result in pregnancy under normal conditions of sexual intercourse."

- The World Health Organization defines male fertility as "the ability of a man to achieve a pregnancy in a fertile female partner."

- The National Institute of Child Health and Human Development defines male fertility as "the ability of a man to cause pregnancy in a fertile female partner."

- In animals, the term fertility is being used to describe the ability of domestic animals to reproduce and produce viable offspring.

- In general, male fertility is defined as the ability of male animals to produce healthy and viable sperm that can fertilize female animals and result in the production of viable offspring.

All mentioned definitions emphasize the basic theme that male fertility is the ability of a male to contribute to a successful pregnancy in a fertile female of the same species.

1.2 What is male infertility? In light of the organizational perspective

Deviation from the normal male fertility process leads to the abnormal state known as "infertility" or "subfertility." However, different tangible and intangible factors could influence the condition. Infertility could also be defined in many ways in humans and animals. Here are a few different definitions of male infertility:

- "The inability of a man to achieve a pregnancy in a fertile female after 12 months or more of regular unprotected sexual intercourse" (The World Health Organization).
- "The inability of a man to achieve a pregnancy in a fertile female after one year of regular unprotected intercourse, or the inability to produce a pregnancy after surgical, medical, or behavioral intervention" (The American Society for Reproductive Medicine).
- "The inability of a man to cause pregnancy in a fertile female partner" (The National Institute of Child Health and Human Development).
- "The inability of a man to achieve conception or to induce conception in a fertile female partner due to quantitative and/or qualitative deficiencies of his spermatozoa" (The European Society of Human Reproduction and Embryology).

In the case of animals, it is defined as:

- "The inability of a male animal to produce viable offspring due to abnormalities or dysfunction in its reproductive system."

These definitions highlight a condition, where a male cannot contribute to a successful pregnancy with a fertile female partner. However, the variations in terms might be linked to the diagnosis process, the specific criteria for infertility, or other factors that result in male infertility.

2. Oxidation-reduction process in males

Oxidation-reduction reactions, also known as redox reactions, which are key determinants in male reproduction, alike many biological processes. Generally, redox reactions involve the transfer of electrons between molecules and result in the molecules' oxidation state.

The redox reactions are involved in male reproduction through:

- Regulation of hormone levels
- Production of quality sperm
- Regulation of sperm maturation
- Promoting acrosome reaction for fertilization
- Prevent the sperm DNA
- Maintain the sperm mitochondrial potential

During sperm maturation, glucose-6-phosphate dehydrogenase protects the developing sperm from oxidative stress. This enzyme reduces the reactive oxygen species (ROS) by catalyzing the first step in the pentose phosphate pathway to produce a reducing agent, i.e., NADPH. In mature sperm, glutathione plays a vital role as an antioxidant and protects sperm cells from oxidative stress by forming the disulfide bonds in the tail region of sperm for motility.

2.1 Oxidative stress: story in male infertility

The oxidative stress is an imbalance between the production of ROS (by cellular respiration and catalytic responses of cytochrome steroidogenic enzymes P450) and the body's antioxidant levels. It reacts and affects the lipids, proteins, and DNA that affect the biological process and lead to cellular affections. Oxidative stress is involved in the pathogenesis of diseases and the non-pathogenesis of clinical conditions. ROS production results from metabolic rate, polymorphisms or mutations in mitochondrial and nuclear DNA, lowered antioxidant production and repair rate, and metal ions and other toxins that affect the oxidation process [8].

Oxidative stress has been defined by different organizations as follows:

- "An imbalance between the production of free radicals (reactive oxygen and nitrogen species) and the ability of the body to counteract or detoxify their harmful effects through neutralization by antioxidants" (National Institute of Environmental Health Sciences).

- "A disturbance in the pro-oxidant-antioxidant balance in favor of the former, leading to potential damage" (The International Union of Biochemistry and Molecular Biology).

- "A condition that arises when the rate of production of reactive oxygen species (ROS) exceeds the rate of their elimination by antioxidant defenses, resulting in damage to cellular macromolecules and organelles" (The American Society for Biochemistry and Molecular Biology).

2.2 Oxidative stress and male behavior

2.2.1 Erectile dysfunction

The oxidative stress alters the morphology of blood vessels and associated nerves for erectile function, which further decreases the bioavailability of nitric oxide (NO) and induces mitochondrial dysfunction, inflammation, and endothelial NO synthase uncoupling. These alterations develop endothelial dysfunction by increasing the adhesion of monocytes to endothelial cells, impairing the angiogenic potential of endothelial cells, and inducing apoptosis. In addition, ROS-induced mitochondrial dysfunction contributes to the additional ROS production by altering mitochondrial metabolism, leading to the exacerbation of endothelial dysfunction [9, 10].

Oxidative stress has been linked with erectile dysfunction due to the excessive generation of free radicals in the cavernosal tissues. Superoxide combines with NO to form highly toxic peroxynitrite that induces lipid peroxidation and damages endothelial cells of penile vessels and cavernosal tissues. Recent trends in the management of erectile dysfunction involve increased NO levels using arginase inhibitors. Because in erectile dysfunction, there are elevated levels of arginase activity, which limits NO synthase activity, reduces NO biosynthesis, and increases arginine degradation [11]. The upregulation of the RhoA/ROCK pathway under oxidative stress damages corpus cavernosum smooth muscle function resulting in erectile dysfunction [12].

2.2.2 Reduced libido

During the process of steroidogenesis and spermatogenesis, normal level of ROS is produced by mitochondrial respiration and catalytic responses of cytochrome steroidogenic enzymes P450. The systemic hormones (FSH, LH, testosterone, E2, PRL) also regulate the total antioxidant capacity in animals. Testosterone and melatonin have a strong antioxidant capacity and protect sperm and testosterone-producing cells against damage by ROS. Damages to Leydig cells and the hypothalamic-pituitary-gonadal (HPG) axis due to excessive ROS decreases testosterone production. However, under oxidative stress condition, the hypothalamic-pituitary-adrenal (HPA) and hypothalamic-pituitary-thyroid (HPT) axes are activated by the excessive production of ROS and stimulate the cortisol hormone and down-regulate the T3 and T4 synthesis. This hormone negatively affects the anterior pituitary gland through crosstalk between HPG and HPA axes and, finally, decreases the secretion of LH and FSH hormones. This lower level of LH, it is not enough to stimulate Leydig cells for testosterone production. This halted testosterone production altering the sialic acid levels in Sertoli and Leydig cells, decreasing the release of androgen-binding protein (ABP) to Sertoli cells, reducing T3 decreases StAR mRNA levels in Leydig cells, and increasing aromatase activity for testicular E2 hormones. Moreover, the ROS-induced alteration in Sertoli cells is associated with reduced hormonal secretions, blockage of gap junction communications, and damage to the blood-testis barrier. Increased oxidative stress also stimulates proinflammatory cytokines due to genital tract infection (tumor necrosis factor-α, interleukin-1β, and interleukin-6), which further harms the HPG axis and testosterone biosynthesis. ROS under obese conditions forcibly drives the adipocytes to secrete leptin along with insulin, negatively affecting the T3 and testosterone synthesis. Hence, ROS leads to male infertility by disrupting reproductive endocrine glands [8], and lower testosterone level "main regulator reproductive behavior" may suppress sexual activities in male.

2.2.3 Impaired sperm transport

Epididymis has an important role in sperm transportation, maturation, and storage. In addition, the epididymis provides essential proteins to spermatozoa *via* epididymosomes to maintain cellular functions and protect them from potential damages such as oxidative stress-dependent injuries. The peroxiredoxins (PRDXs) are a family of antioxidant enzymes highly expressed in yeast to humans. They are peroxidases that do not require cofactors such as heme group or selenium and contain one or two cysteines (Cys) residues in their active site, which are essential for their antioxidant function. PRDXs isoforms are divided into 2-Cys PRDXs (PRDX1-4), atypical PRDX (PRDX5), and 1-Cys PRDX (PRDX6). These enzymes are important antioxidants in spermatozoa that regulate the level of ROS, such as peroxides (H_2O_2 and organic hydroperoxides) and peroxynitrite (ONOO−), to avoid cellular toxicity [13].

The metabolic processes occurring within the epididymal tissue and the absorptive and secretory activities of epididymal epithelium, which provide an optimum internal milieu for sperm maturation, are regulated by androgens. Androgen deprivation leading to apoptosis in epididymal epithelium has been demonstrated [14]. Sialic acid, secreted by the principle cell, not only confers a negative charge to the sperm surface but also acts as a decapacitation factor protecting the spermatozoa from premature capacitation. Hence, sialic acid is essential for maturation and maintenance of plasma membrane structural integrity of spermatozoa in epididymis [15].

2.3 Oxidative stress and sperm production

Notwithstanding the antioxidant protection afforded to the testes to support its dual functions of steroidogenesis and sperm production, a wide variety of endogenous and exogenous factors are known to perturb these defenses and generate oxidative stress. Here are a few ways that oxidative stress can affect sperm production:

2.3.1 Reduced spermatogenesis

The diet, alcohol consumption, less physical activity, heat, radiation, chemical exposure, pollution, smoking, varicocele operation, and trauma influence sperm parameters and impair spermatogenesis [16]. Such changes are mostly linked to the high susceptibility of the unsaturated fatty acid content of sperm to oxidative stress. The lipid oxidation initiates the ROS that attacks the double bonds of an unsaturated fatty acid of the plasma membrane. Firstly, the ROS reacts with the adjacent lipid molecule triggering a chain reaction. By-products of lipid oxidation include mutagenic and genotoxic molecules malondialdehyde (MDA) and 4 hydroxy-nonenal (4-HNE) that cause the DNA damage. Secondly, DNA damage occurs due to free radical-initiated apoptosis, leading to caspase-mediated destruction of DNA [17, 18]. The molecular aspects in male germ cells, the epigenetic modification of the imprinting of paternal genes, DNA compaction, and silencing of the post-meiotic gene are necessary steps for spermatogenesis. However, altered methylation and acetylation of histone proteins H3 and H4 at specific lysine residues impair spermatogenesis [19].

2.3.2 Increased germ cell apoptosis

In general, high levels of ROS production induce apoptosis in the germ cells and clear the nonviable sperm from the testes. Under pathological conditions, a massive germ cell apoptosis affects the spermatocytes frequently, spermatogonia minutely, and spermatids rarely. The Sertoli cells are less prone to apoptosis as compared to the germ cells. The *Fas*-system, which is expressed in Sertoli cells, is responsible catering apoptosis in germ cells. In healthy males, the Sertoli cells express *FasL*, which triggers the apoptosis in *Fas*-positive germ cells in a paracrine manner among germ and Sertoli cells. The Bcl-2 (Bcl-2 and Bax) and caspase (caspase-3/-8/-9/-12) family members control the mitochondrial apoptosis process. An imbalance of the Bax and Bcl-2 ratio and mitochondrial potential depolarization activates the caspase-9 process and triggers apoptosis [20, 21].

2.3.3 Loss of Leydig cell function

The production of testosterone is major function of the Leydig cell, and is affected by several ways under increased ROS production. Firstly, the ROS affects the Leydig cell function by reacting with Leydig cell DNA and oxidize the DNA leads to DNA strands breakage and mutations. Secondly, Increased ROS is involved in oxidation of unsaturated fatty acids of Leydig cell plasma membrane, which further promote the ROS production. Thirdly, the depletion of antioxidants made the Leydig cells more prone to accumulated higher ROS levels. Lastly, provoking the release of cytokines by increased ROS levels influence the Leydig cell physiology. Under low testosterone conditions, the Leydig cells' activity is hampered, and apoptosis occurs by stimulating caspase activity and DNA damage of Leydig cells. Altered testosterone and estradiol levels affect the availability of LH receptors on Leydig cells, changing Sertoli cells' functionality for germ cell survival, and seminiferous tubular maturity [21].

2.3.4 Impaired testicular blood flow

Like other deleterious consequences of oxidative stress to the vascular system, impairment of testicular blood vessels minimizes the blood flow and oxygen delivery to the testes, and negatively impacting sperm production. Under pathological conditions of testicular torsion, testicular ischemia develops, and hence oxidative stress occurs in the testicles of the same side. Then the NO and H_2O_2 production increases leading to lipids peroxidation. The accumulation of isoprostane and decrease in antioxidant enzymes increase the apoptosis. Short periods of ischaemic conditions can also develop testicular oxidative stress, decreasing testicular antioxidants with reduced spermatogenesis [22]. During testicular varicose, endothelial NO synthase (eNOS) production increased, and blood flow toward testicular tissues also increased to compensate for the hypoxic condition due to venous stagnation. Increased NO concentrations react with superoxide free radicals and produce the reactive nitrogen species (peroxynitrite and peroxynitrous acid) that impair fertility. Moreover, increased leptin receptors (glial cell line-derived neurotrophic factor receptor-1 and voltage-dependent calcium channels) are also predisposing factors for oxidative stress under testicular varicose [23].

2.4 Oxidative stress and epididymis function

Compromised epididymal functions (removal of excess cytoplasm and acquiring motility and fertilization potential) due to oxidative stress is a leading cause of male infertility. The sperm maturation, storage, and transportation are down-expressed under oxidative stress conditions because epididymis is highly sensitive to oxidative stress due to the high levels of polyunsaturated fatty acids in its membranes and the presence of ROS-generating enzymes. The following lines show how oxidative stress can affect epididymis function:

2.4.1 Impairment in sperm maturation

In the epididymis, increased expression of antioxidant enzymes reduces oxidative damage to sperm through the secretion of epididymosomes. The high 4-HNE levels in caput and cauda epididymis result in lipid peroxidation and epididymal epithelium degeneration by oxidative stress and hamper sperm maturation. The increased production of PRDX6 and PRDX1 seems insufficient under oxidative stress to quench excessive ROS and maintain a conducive cellular environment for sperm maturation [13]. Increased production of immature sperm and increased ROS levels that severely affect epithelial cells of the epididymis. High ROS generation or failure of antioxidant systems provoke the eukaryotic cells to combat the deleterious by-products of redox that result in the oxidative injury of sperm.

2.4.2 Sperm DNA damage

The sperm at the testicular level are highly prone to ROS and induce DNA fragmentation in sperm. DNA fragmentation is relatively low during epididymal transit due to high ROS production. In the epididymis environment, sperm nuclear condensation and oxidation are interlinked processes. The DNA compaction is completed by creating inter- and intramolecular cross-links between nuclear protamines. For bridging sperm protamines, a balanced epididymis ROS and enzymes (protein disulfide isomerase and glutathione peroxidase) are required. Excessive ROS generation by immature sperm in the epididymis not only affects the sperm membranes but also multiplies ROS production that transiently increases sperm nuclear condensation and quickly reverses the DNA fragmentation due to nuclear decondensation and oxidative-induced DNA fragmentation [24].

This oxidative damage increases DNA fragmentation and lipid peroxidation in epididymal cells. In the epididymis, ROS oxidizes the guanosine to 8-deoxy-2′hydroxy-guanosine up-regulates the DNA mutation and damage to the paternal genome, leading to male infertility [25]. The disruption of molecular mechanisms driven by miRNAs or epigenetic changes can induce permanent sperm DNA damage in the epididymal climate [13].

2.4.3 Disrupted epididymal fluid composition

Oxidative stress alters the epididymal fluid composition and negatively impacts sperm function and viability. Induction of oxidative stress creates a disturbance of epididymal physiology, alterations in ion and fluid transporters within the epididymis, and abnormal water reabsorption result in significant changes in the luminal fluid composition [14, 26].

Obesity change in the epididymis's microenvironment by increasing MDA expression in epididymal fluids with low glutathione levels that impair sperm quality. Epididymosomes secreted in the epididymal lumen communicates with the sperm by sharing the protein and noncoding RNA contents, lowering oxidative stress levels [27].

2.4.4 Inflammation in epididymis

The inflammation in the epididymis induces ROS that damages the tissue and impairs epididymal function. Although the presence of proinflammatory cytokines, tumor necrosis factor-alpha (TNF-α), interleukin-1 alpha (IL-1α), and interleukin-1 beta (IL-1β) cytokines in the epididymis are involved in certain physiological events, the increased TNF levels influence sperm motility, viability, and DNA fragmentation [28, 29].

2.5 Oxidative stress and accessory sex glands

Accessory sex glands (prostate gland, seminal vesicles, and bulbourethral gland glands) secrete the seminal plasma of the ejaculate that provides nutrients and protection to sperms. The seminal plasma contains enzymatic (superoxide dismutase; SOD, catalase; CAT, glutathione peroxidase; GPx) and nonenzymatic antioxidants (vitamins C and E, hypotaurine, taurine, uric acid, albumin, etc.). Among these enzymes, the SOD predominantly protects sperm against oxidative stress induced by NADPH treatment [30]. Oxidative stress affects the function of the prostate gland, seminal vesicles, and bulbourethral gland glands. The following lines indicate the deleterious effects of oxidative stress on accessory sex glands:

2.5.1 Altered seminal fluid production and composition

Increased ROS level damage the cells of accessory sex glands and affecting the production and composition of seminal fluid. Altered pH, ion concentration, and nutrient content in response to increased ROS negatively affect sperm function and viability. Seminal fluid hyperviscosity due to oxidative stress leads to low seminal plasma fructose, ascorbic acid, calcium, and zinc levels, which indicate low glandular activity. Changes in prostasomes (cholesterol, sphingomyelin, calcium, and different enzymes) due to high ROS generation reflect the low prostate gland activity [31]. The acidic pH of seminal fluid indicates high ROS, reduced sperm motility, and viability that is neutralized by bicarbonate produced by accessory sex glands. The bicarbonate levels in seminal plasma are regulated by the enzymatic activity of cytosolic carbonic anhydrase isoenzymes (CA-I, CA-II, and CA-III) in accessory sex glands [32].

2.5.2 Impaired glandular contraction

Increased ROS influences the contractility of the smooth muscle cells in the accessory sex glands that, in turn, reduce the seminal fluid volume at ejaculation time. However, anejaculation corresponds primarily to a lack of seminal fluid emission that may be caused by impaired seminal vesicle contraction and seminal fluid production. Alteration in the expression α1- adrenergic receptors in response to high ROS generation also affects the contraction of seminal vesicles and vas deferens [33]. Nerve dysfunction due to oxidative stress influences the sympathetic neuronal input, affecting the secretory pattern of the ductus ejaculatory closure resistance and contraction of the seminal vesicles smooth muscle cells [34]. In addition, cytotoxicity by oxidative stress reduces the contractility of the smooth muscle of accessory sex glands by expressing calcium channel blockers [35].

2.5.3 Increased inflammation

Oxidative stress can also induce inflammation in the accessory sex glands, damaging tissue, and impairing glandular function. Infiltration of leukocytes in infectious conditions provokes ROS overproduction. In response, the generation of proinflammatory cytokines (IL-1b, IL-6, IL-8, IL-10, and IL-12) modulates the activity of pro- and anti-oxidative systems, whereas the oxidative stress exerts its effects even at the end of infection that impair the sperm function [31, 36].

2.6 Oxidative stress and sperm function

Oxidative stress greatly affects sperm motility, morphology, and fertilization potential. The following lines reflect the effects of oxidative stress on motility, concentration, morphology, acrosome reaction, DNA, mitochondrial potential, etc.

2.6.1 Reduced sperm motility

Deviation in motility pattern coincides with the ROS-induced damages in the axonemal part of sperm, which are linked to the following mechanisms:

- Increased susceptibility of PUFAs in the plasma membrane of sperm leads to lipid peroxidation. This oxidative stress induces axonemal damage and reduced sperm viability with higher midpiece morphological defects, which affect sperm motility [37].

- The lower activity of G6PD along high cytokines levels is also one of the major causes of decreased sperm motility, which is indicative of oxidative stress in the seminal plasma due to increased MDA [38].

- An increase in S-glutathionylation and tyrosine nitration of sperm proteins by oxidative stress deleteriously affects sperm motility [39].

- Activation of mitochondrial oxidative phosphorylation in response to oxidative stress is involved in bioenergetic pathways for low sperm motility [40].

- The 4HNE influences the activity of metabolic enzymes in sperm, which provide the energy for sperm motility [41].

2.6.2 Altered sperm morphology

Increased ROS production leads to structural abnormalities in sperm, which halt the fertilization potential. Immature sperm or disrupted sperm morphology and cytoplasmic droplets are prime factors for increased ROS generation. The defected cytoplasm further stimulates abnormal sperms to produce endogenous ROS by activating the enzyme glucose-6-phosphate dehydrogenase [21]. In addition, oxidative stress in the male reproductive tract promotes apoptosis, elevating morphological defects in developing germ cells [42].

2.6.3 Reduced sperm count

Oxidative stress also affects sperm concentration in ejaculated semen, but the extent of the effect is not as great as seen in other sperm attributes. However, low sperm count is observed when males are exposed to prolonged exposure to oxidative during disease conditions that affect the seminiferous epithelium by elevated ROS production. Under such conditions, testicular atrophy is an indication due to ROS-induced damage in the seminiferous tubules [37].

2.6.4 Impaired acrosome reaction

Oxidative stress impairs the capacitation process and disrupts the acrosome integrity in sperm, which reduces the fertilization potential. Under normal circumstances, ROS facilitates the capacitation process by activating intracellular cAMP by downstream PKA that phosphorylates MEK (extracellular signal-regulated kinase)-like proteins, threonine-glutamate-tyrosine, and fibrous sheath proteins. This mechanism further promotes the capacitated sperm for the acrosome reaction [43]. To ensure fertilization, the hyperactive sperm cross the cumulus oophorous, bind to the oocyte, and create a pore in its extracellular matrix *via* the exocytotic release of proteolytic enzymes. These acrosome reactions are mediated through the phosphorylation of tyrosine proteins and Ca^{+2} influx resulting in an intracellular rise in cAMP and PKA, enabling the spermatozoon to penetrate and fuse with the oocyte. ROS has been observed to facilitate actions on the zona pellucida of the spermatozoon by various means, including phosphorylation of three relevant plasma membrane proteins [44].

2.6.5 Sperm DNA fragmentation

ROS affects the sperm DNA integrity by base modification, telomere shortening, and epigenetic changes that interfere with DNA replication and transcription, chromosomal instability and genetic abnormalities in the sperm, and altered gene expression. ROS damage the sperm DNA by attacking the purine and pyrimidine bases, causing single and double-strand DNA breaks, cross-linkage, or chromosomal rearrangements. Moreover, deficiency in sperm protamination leads sperm DNA more prone to ROS effects. Secondly, ROS-initiated apoptosis damages sperm DNA by caspase-mediated DNA destruction. In ROS-induced sperm DNA damage, 8-hydroxy-2′-deoxyguanosine (8-OHdG) is the potential indicator of DNA damage. The ensuing oxidative stress drives the spermatozoa along the intrinsic apoptotic cascade, from loss of MMP to oxidative DNA adduct formation, DNA fragmentation, and ultimately, cell death [45].

2.6.6 Loss of sperm mitochondrial membrane potential (MMP)

Oxidative stress also significantly impacts the mitochondria in sperm, affecting sperm motility and function. Here are a few ways that oxidative stress can affect sperm mitochondria:

- Increased levels of polyunsaturated fatty acids (PUFAs) initiate sperm mitochondrial ROS production, which occurs due to the dual hydrophilic and hydrophobic properties of PUFA. Penetration of the inner mitochondrial membrane disrupts the ETC electron flow, which produces the superoxide anions; subsequently, oxidative stress develops [46].

- Oxidative stress in the sperm also induces the generation of mitochondrial ROS. Adducts of aldehydes, by-products of lipid peroxidation (e.g., 4-hydroxynonenal), and mitochondrial proteins within the electron transport chain (ETC) further stimulate the mitochondrial ROS production [39, 47].

- Increased mitochondrial ROS levels induce oxidative stress by promoting apoptotic activity through a truncated apoptotic pathway [46].

- Increased mitochondrial permeability increases the intracellular Ca^{2+} that loss of MMP by lowering the ATP content and increased ROS, which deteriorates the plasma membrane integrity [48].

Spontaneous mitochondrial ROS production also results in loss of MMP, increased lipid peroxidation, and impaired sperm motility [49], which is linked to disruption of the mitochondrial electron transport, formation of adducts with mitochondrial proteins, reduced mitochondrial expression of prohibitin, the opening of the mitochondrial permeability transition pore, and induction of apoptosis in exposed sperm (reviewed in Aitken [46]).

In addition, nitric oxide radicals (by diffusion or generated within the mitochondria) react with superoxide anions (derived from the inner mitochondrial membrane) to form the peroxynitrite, which further produces adduct, that is. 4-hydroxynonenal protein adducts. Formation of such adducts leads to reduced motility *via* mitochondrial dysfunction [50–54].

2.6.7 Damaged sperm plasma membrane

Oxidative stress affects the membrane of sperm through protein or lipid peroxidation, reducing membrane fluidity. The sperm plasma membrane is crucial in sperm motility, capacitation and fertilization. Here are a few ways that oxidative stress can affect the sperm membrane:

- High contents of PUFA in the sperm plasma membrane make its susceptibility to membrane lipid peroxidation. However, PUFAs are necessary for membrane fluidity and fusogenicity of sperm membranes because the lowest carbon-hydrogen dissociation energies at the bisallylic methylene position in PUFA are easy targets of ROS [52].

- PUFA-rich sperm plasma membrane regulates membrane fluidity and is a target site of ROS to promote lipid peroxidation cascade, which decreases membrane fluidity [51].

- ROS produced by the NADPH oxidase system also induces damage in the sperm plasma membrane [55].

3. Nutraceuticals

Nutraceuticals are categorized as food or components used for health benefits beyond basic nutrition. Nutraceuticals could be in dietary supplement forms, functional foods (fortified or enhanced with additional nutrients), or bioactive compounds. Nutraceuticals are present in a single or in combination with vitamins, minerals, herbal supplements, probiotics, prebiotics, and omega-3 fatty acids. Mostly, nutraceutical remedies reduce inflammation, boost immune function, improve cardiovascular health, and promote healthy aging.

The popularity of nutraceuticals has been boosted in recent years on the eve of global antimicrobial resistance, and scientists are moving toward alternative approaches to assure health and well-being. However, it is noteworthy that many nutraceuticals possess promising health benefits, but scientific evidence about their safety and mechanism is limited. Similarly, the use of nutraceuticals to improve sexual health is also widespread under physiological or pathological infertility. Nutraceuticals to improve male infertility against oxidative stress have been extensively used and partially followed the plausible involved mechanisms.

3.1 Use of nutraceuticals against male infertility: underlying mechanisms

While there is limited research specifically examining the effects of nutraceuticals on copulation, there is some evidence to suggest that certain nutraceuticals may positively impact sexual function and libido in both men and women. Some examples of nutraceuticals that have been studied in this context include:

3.1.1 Nutraceuticals and erectile dysfunction

- **L-arginine** is an amino acid responsible for nitric oxide in the biological cycle that involves dilating penile vessels for blood flow to enhance penile erection. Arginase enzyme regulates NO synthase generation in corpus cavernosal tissues, leading to penile erection. Under erectile dysfunction, high arginase activity diminishes the arginine level. It is not available for synthesis NO, the substrate of endothelial NO synthase and neuronal NO synthase [56] for penile erection by provoking the relaxation of vascular and nonvascular cavernosal tissues *via* cGMP activation [57]. L-arginine promotes the NO and cyclic guanosine monophosphate production for erectile function. Under oxidative stress, arginine is converted to NO, which also scavenges the excessive ROS from penile tissues and improves erectile function [58, 59].

- **Horny goat weed** is a nutraceutical, which is being widely used for erectile dysfunction and possesses the active bioactive ingredient icariin. Icariin has the potency of a PDE5 inhibitor and boosting properties of testosterone. Moreover, icariin increases smooth muscle proliferation and has neurotrophic effects, which are useful for erectile dysfunction due to oxidative stress-induced endothelial cell damage [60].

- In addition, **maca root** is used as an aphrodisiac. Although the exact mechanism of action how it mitigate oxidative stress and improves penile erection is not clear, the antioxidant properties could be speculated about it [61].

- **Muira Puama** and **Saw Palmetto** are potential nutraceuticals that improve erectile dysfunction by inhibiting PDE5 activity and quenching excessive ROS in penile tissues. Administration of maca root is a choice supplement in case of erectile dysfunction by active ingredients of maca roots, i.e. macamides and macaenes, etc., which potentially boost the antioxidant level in male reproduction. Moreover, the erectile-enhancing effect of maca roots is linked to glucosinolate extracts that activate the androgen binding receptors in penile tissues [62]. However, the particular mechanism of maca root for erectile dysfunction is still unexplored.

- **Ginseng** contains potential aphrodisiacs because of active agents (ginsenosides and ginseng saponins). These compounds are observed to increase nitric oxide (NO) synthase activity, increasing the blood flow to the penile corpora cavernosa. The Antioxidants property of ginseng is also the mechanism to combat erectile dysfunction [63].

3.1.2 Nutraceuticals and male libido

The effect of each nutraceutical is linked to various potential mechanisms to improve the libido.

- The use of **Korean red ginseng** is established in boosting sex hormones by providing an environment of antioxidants in the testes because of the enzymatic and nonenzymatic characteristics of ginsenosides [64].

- The **Ashwagandha** (*Withania somnifera*) improves libido through strong cellular antioxidant effects. It reduces the ROS level by improving the metal cofactors of SOD, and catalase has a key role in steroidogenesis. Moreover, it restores the sex hormone level by promoting antioxidant vitamins and minimizing cortisol under stressful conditions [65]. Another involved mechanism of *W. Somniferous* for improving sexual hormones *via* activating the Nrf2/HO-1 pathway and inhibiting the NF-κB levels is explained [66].

- *Tribulus terrestris* improves the libido level by improving the testosterone level. It directly increases the testosterone, dihydrotestosterone, and dehydroepiandrostenedione, which improve libido. It is also linked to the proliferation of testicular cells through the increased conversion of testosterone into dihydrotestosterone by the action of 5-α reductase [67].

- Tongkat Ali's also improves the libido, but the mechanism of its action for increased testosterone is unclear.

- Supplemental **zinc** improves hypogonadism by boosting testosterone levels due to its potential action of antioxidant and cofactor of the enzyme involved in steroidogenesis [68].

- **Fenugreek seed extract**, composed of numerous enzymes, amino acids (including arginine), vitamins, and lipids, improve the libido by maintaining the glucose and cholesterol level for steroidogenesis [69].

3.1.3 Nutraceuticals and spermatogenesis

- **Vitamin D** regulates the testicular functions by enabling the Sertoli, germ cells, Leyding cells, and epithelial cells lining the male reproductive tract to utilize the calcium for metabolism [70].

- **Vitamin E** is an antioxidant that may help to protect testicular cells from oxidative damage. It consists of biologically active compounds of tocopherols and tocotrienols that scavenge the ROS and prevent spermatogonium degeneration, testicular dysfunction, and seminiferous tubule shrinkage. The antioxidant activities of superoxide dismutase (SOD) and glutathione peroxidase (GSH-PX) in testicular cell membranes and mitochondria are promoted by vitamin E integration. Along with boosting testosterone levels, Zinc is involved in anatomical configuration and normal functioning of male reproductive organs, sperm maturation during spermatogenesis and preventing the germinative epithelium from ROS affections [6].

- **Coenzyme Q10** (CoQ10) is an antioxidant involved in energy production in the body, testicular function, and testosterone production. CoQ10 supplementation affects spermatogenesis by lowering FSH levels and increasing inhibin B levels [71–74].

- **Ashwagandha** contains numerous bioactive compounds, which improve spermatogenesis due to its antioxidative properties. In addition, its supplementation ensures the availability of different metabolites, that is. iron, alanine, aspartate, fructose, lactate, glutamine, etc., for testicular functions [75, 76].

- Beyond the antioxidant or aphrodisiac effects, **Tongkat Ali** is involved in minimizing the inhibitory effects of estrogen on spermatogenic cells during spermatogenesis [77].

- **Korean red ginseng** improves spermatogenesis through testicular antioxidants homeostasis (glutathione, vitamin C and E, and expressions of phosphatidylinositol transfer protein, fatty acid-binding protein-9, and triosephosphate isomerase-1 proteins) and ginsenosides (dammarane-type triterpene saponin) effects on re-establishing of sperm maturation process [78].

3.1.4 Nutraceuticals and sperm motility

- **Carnitine** is an amino acid regulating intracellular metabolism through b-oxidation and buffers the acetyl-coenzyme A (CoA) to CoA ratio by transporting long-chain fatty acids into the mitochondria. This process is involved in energy production in sperm for sperm motility and maturation. Moreover, the antioxidant property of carnitine protects the sperm against excessive ROS and maintains motility [79].

- **Vitamin C** is an antioxidant that improves sperm motility by reducing oxidative stress. It is a key cofactor for hydroxylation and amidation reactions and is involved in the synthesis of collagen, components of the intercellular matrix, and proteoglycans for cellular metabolism. Its antioxidant property prevents ROS attack on sperm plasma and mitochondrial membrane and maintains motility [80].

- **CoQ10** has strong antioxidant potential and plays a role in cell energy production. Improvement in sperm motility by CoQ10 is maintained by providing energy. Mitochondrial bioenergetics is known for sperm motility and requires a high energy expenditure, which is fulfilled by CoQ10 when oxidative stress exists [74, 81].

- **Omega-3 fatty acids** are important for reproductive health. Repair of plasma and mitochondrial membrane and lipid metabolism by Omega-3 fatty acids supplementation under oxidative stress provides sperm membrane integrity that, in turn, is helpful for sperm motility [82]. Promotion of the sperm lactate dehydrogenase isoenzymatic form by omega-3 PUFA supplements maintains the catalytic conversion of pyruvate to lactate in the energy metabolism of sperm [83].

- **Zinc** is an essential mineral that is important for male reproductive health. It helps to increase sperm motility. Zinc is an enzyme cofactor for DNA transcription and protein synthesis that might improve sperm metabolism and maintain motility [84, 85].

3.1.5 Nutraceuticals and sperm concentration

- **CoQ10** supplements regarding sperm count, lower FSH, and higher inhibin B enhance the Sertoli cell function, improving sperm concentration [86].

- **Vitamin B12** is involved in sperm metabolic activity. It is a coenzyme that reduces ribonucleotides to deoxyribonucleotides. It also stimulates growth and maintains synthesis and sperm maturation. It prevents the methylmalonic acid that increases ROS-induced incidences in sperm [87].

- **Zinc** plays a role in testicular development and spermatogenesis. It has a role in DNA function regulation in sperm cells and indirectly improves spermatogenesis by promoting testosterone production. Moreover, the antioxidant effect of Zn protects the Leydig cells from ROS-induced damage and maintains normal sperm concentration [88, 89].

- **L-acetyl carnitine** protects sperm mitochondria from free radicals. It stabilizes spermatogonial production through antiapoptotic action and maintains sperm count under oxidative stress [90, 91].

3.1.6 Nutraceuticals and sperm morphology

- **Omega-3 fatty acids** possess the α-linolenic acid (ALA), docosahexaenoic acid (DHA), and eicosapentaenoic acid (EPA). The supplementation of these under oxidative stress stabilizes the cell membrane composition by incorporation into the sperm cell membrane and maintains the sperm morphology [83, 89].

- **L-carnitine** is an amino acid that is important for energy production in cells. It may help to improve sperm morphology by providing energy to the sperm cells. Irrational incidence of apoptosis during spermatogenesis due to excessive ROS leads to sperm morphological abnormalities, which is ameliorated by L-carnitine supplement. It seems that the antioxidant mechanism of L-carnitine is involved in reducing sperm morphological defects [90, 92].

- **Vitamin D**, a membrane-bound antioxidant, has a potential scavenger capacity. Oxidative stress marker (4-hydroxynonenal) is the main indicator of vitamin D deficiency which increases sperm morphological defects; hence, vitamin D supplement and expression of vitamin D receptors in the male reproductive tract could minimize sperm morphological defects [93].

- **Zinc** uptake can minimize oxidative stress and maintain Leydig cell integrity and steroid hormone synthesis, which improves sperm morphology [88].

- **CoQ10** supplementation improves sperm morphology by providing energy to the sperm cells. Moreover, it transports electrons in the mitochondrial respiratory chain to promote energy in the mid-piece region of sperm. It also acts as a lipid-soluble antioxidant for the lipoprotein-rich cell membrane and stabilizes sperm morphology against oxidative stress [74].

3.1.7 Nutraceuticals and sperm mitochondrial integrity

- **CoQ10** supplement mitigates the excessive ROS production during the mitochondrial oxidative phosphorylation process due to its antioxidant potential. Promoting the mitochondrial electron transport chain by CoQ10 preserves ATP production and mitochondrial transcription, maintaining sperm motility [94].

- **L-carnitine** is involved in energy metabolism within sperm mitochondria. L-carnitine supplements improve motility by protecting sperm mitochondria against ROS [95].

- **Omega-3 fatty acids**, docosahexaenoic acid (DHA), and eicosapentaenoic acid (EPA) possess antioxidant effects, which reduce oxidative stress in sperm mitochondria and improve sperm motility. PUFA supplementation reduces the excessive ROS by maintaining the βeta-oxidation in sperm mitochondria for sustainable sperm functions [96].

- **Myoinositol** is a component of the vitamin B complex and maintains sperm mitochondrial potential for motility. Myoinositol might increase cytosolic Ca^{+2} in sperm, which also increases mitochondrial Ca^{+2} necessary for mitochondrial potential and sperm motility [94].

3.1.8 Nutraceuticals and sperm DNA

- **Vitamin C** mitigates the excessive ROS in sperm and reduces the disulfide bridges of cysteine residues and improves the testicular antioxidants, which protect the sperm from chromatin damage [97].

- **Zinc** is an integral element in DNA synthesis. Zn is also required for correct sperm DNA condensation/decondensation. Chromatin stability of the ejaculated sperm is Zn-regulated and controls the disulfide bridge formation [89, 98].

- **CoQ10** is a strong antioxidant, and its reduced form, ubiquinol, and prevents oxidative stress in sperm DNA [73, 74, 99].

- **L-Carnitine** is a powerful antioxidant and reduces sperm DNA decondensation by scavenging O_2^- and H_2O_2 and inhibiting iron-mediated ROS production [97, 100].

- **Folic acid** is intrinsically involved in purine and pyrimidine production that plays an important role in DNA synthesis and proper sperm function. It is also a potent free radical scavenger by providing methyl donors methionine and S-adenosylmethionine, decreasing the frequency of sperm DNA abnormalities [87].

4. Conclusions

The current evidence shows how oxidative stress influences male fertility by impairing the male reproductive tract environment, reducing the hormonal levels, and damaging the testicular or sperm functions. At the same time, numerous biological agents such as minerals, vitamins, and herbs possess single or multiple active compounds that directly improve the enzymatic antioxidant activity or indirectly enhance antioxidant levels through cellular metabolism and scavenge the excessive ROS from the male reproductive system. It is evident how these antioxidants clear the excessive ROS for better libido and penile erection. The involved antioxidative mechanism linked to famous nutraceuticals compounds for spermatogenesis, motility, concentration, morphology, mitochondrial function, and DNA integrity of sperm is separately depicted in this chapter.

In future, use of nutraceuticals in male infertility conditions needs validation by exploring molecular mechanics, long-term effects with high safety profiles, specific pathways, and target sites of nutraceuticals.

Author details

Zahid Naseer[1*], Mudussar Nawaz[1], Ejaz Ahmad[2] and Zia ur Rehman[3]

1 Faculty of Veterinary and Animal Sciences, Department of Clinical Studies, Pir Mehr Ali Shah Arid Agriculture University, Rawalpindi, Pakistan

2 Faculty of Veterinary and Animal Sciences, Department of Clinical Sciences, Bahauddin Zakariya University, Multan, Pakistan

3 Faculty of Animal Husbandry and Veterinary Sciences, College of Veterinary Sciences, The University of Agriculture Peshawar, Peshawar, Pakistan

*Address all correspondence to: zahidnaseer@uaar.edu.pk

References

[1] Mann U, Shiff B, Patel P. Reasons for worldwide decline in male fertility. Current Opinion in Urology. 2020;**30**(3):296-301

[2] D'Angelo S, Meccariello R. Microplastics: A threat for male fertility. International Journal of Environmental Research and Public Health. 2021;**18**(5): 2392

[3] Ritchie C, Ko EY. Oxidative stress in the pathophysiology of male infertility. Andrologia. 2021;**53**:e13581. DOI: 10.1111/and.13581

[4] De Luca MN, Colone M, Gambioli R, Stringaro A, Unfer V. Oxidative stress and male fertility: Role of antioxidants and inositols. Antioxidants. 2021;**10**(8):1283

[5] Ko EY, Sabanegh ES. The role of nutraceuticals in male fertility. Urologic Clinics. 2014;**41**(1):181-193

[6] Beygi Z, Forouhari S, Mahmoudi E, Hayat SM, Nourimand F. Role of oxidative stress and antioxidant supplementation in male fertility. Current Molecular Medicine. 2021;**21**(4):265-282

[7] Kuchakulla M, Soni Y, Patel P, Parekh N, Ramasamy R. A systematic review and evidence-based analysis of ingredients in popular male fertility supplements. Urology. 2020;**136**:133-141

[8] Barati E, Nikzad H, Karimian M. Oxidative stress and male infertility: Current knowledge of pathophysiology and role of antioxidant therapy in disease management. Cellular and Molecular Life Sciences. 2020;**77**:93-113

[9] Agarwal A, Nandipati KC, Sharma RK, Zippe CD, Raina R. Role of oxidative stress in the pathophysiological mechanism of erectile dysfunction. Journal of Andrology. 2006;**27**:335-347

[10] Fujita N, Momota M, Ishida M, Iwane T, Hatakeyama S, Yoneyama T, et al. Association of oxidative stress with erectile dysfunction in community-dwelling men and men on dialysis. The Aging Male. 2022;**25**(1):193-201

[11] Oboh G, Ademiluyi AO, Ademosun AO, Olasehinde TA, Oyeleye SI, Boligon AA, et al. Phenolic extract from Moringa oleifera leaves inhibits key enzymes linked to erectile dysfunction and oxidative stress in rats' penile tissues. Biochemistry Research International. 2015. Article ID 175950. DOI: 10.1155/2015/175950

[12] Yuan P, Ma D, Gao X, Wang J, Li R, Liu Z, et al. Liraglutide ameliorates erectile dysfunction via regulating oxidative stress, the RhoA/ROCK pathway and autophagy in diabetes mellitus. Frontiers in Pharmacology. 2020;**11**:1257

[13] Wu PY, Scarlata E, O'Flaherty C. Long-term adverse effects of oxidative stress on rat epididymis and spermatozoa. Antioxidants. 2020;**9**(2):170

[14] Elbashir S, Magdi Y, Rashed A, Henkel R, Agarwal A. Epididymal contribution to male infertility: An overlooked problem. Andrologia. 2021;**53**(1):e13721

[15] Saradha B, Mathur PP. Induction of oxidative stress by lindane in epididymis of adult male rats. Environmental Toxicology and Pharmacology. 2006;**22**(1):90-96

[16] Aitken RJ, Roman SD. Antioxidant Systems and Oxidative Stress in the

Testes. In: Cheng CY, editor. Molecular Mechanisms in Spermatogenesis. Advances in Experimental Medicine and Biology. New York, NY: Springer; 2009;**636**:154-171. DOI: 10.1007/978-0-387-09597-4_9

[17] Rajender S, Avery K, Agarwal A. Epigenetics, spermatogenesis and male infertility. Mutation Research. 2011;**727**:62-71

[18] Wagner H, Cheng JW, Ko EY. Role of reactive oxygen species in male infertility: An updated review of literature. Arab Journal of Urology. 2018;**16**(1):35-43

[19] Sharma P, Ghanghas P, Kaushal N, Kaur J, Kaur P. Epigenetics and oxidative stress: A twin-edged sword in spermatogenesis. Andrologia. 2019;**51**:e13432

[20] Aitken RJ. Impact of oxidative stress on male and female germ cells: Implications for fertility. Reproduction. 2020;**159**(4):R189-R201

[21] Asadi A, Ghahremani R, Abdolmaleki A, Rajaei F. Role of sperm apoptosis and oxidative stress in male infertility: A narrative review. International Journal of Reproductive Biomedicine. 2021;**19**(6):493-504

[22] Asadi N, Bahmani M, Kheradmand A, Rafieian-Kopaei M. The impact of oxidative stress on testicular function and the role of antioxidants in improving it: A review. Journal of Clinical Diagnosis and Research. 2017;**11**(5):IE01-IE05

[23] Dutta S, Sengupta P, Slama P, Roychoudhury S. Oxidative stress, testicular inflammatory pathways, and male reproduction. International Journal of Molecular Sciences. 2021;**22**(18):10043

[24] Drevet JR, Aitken RJ. Oxidative damage to sperm DNA: Attack and defense. In: Baldi E, Muratori M, editors. Genetic Damage in Human Spermatozoa. Advances in Experimental Medicine and Biology. Cham: Springer; 2019;**1166**:107-117. DOI: 10.1007/978-3-030-21664-1_7

[25] O'Flaherty C. Orchestrating the antioxidant defenses in the epididymis. Andrology. 2019;**7**(5):662-668

[26] Gutiérrez-Reinoso MA, García-Herreros M. Normozoospermic versus teratozoospermic domestic cats: differential testicular volume, sperm morphometry, and subpopulation structure during epididymal maturation. Asian Journal of Andrology. 2016;**18**(6):871-878. DOI: 10.4103/1008-682X.187583

[27] Li Y, Zhao W, Fu R, Ma Z, Hu Y, Liu Y, et al. Endoplasmic reticulum stress increases exosome biogenesis and packaging relevant to sperm maturation in response to oxidative stress in obese mice. Reproductive Biology and Endocrinology. 2022;**20**(1):161

[28] Avellar MCW, Ribeiro CM, Dias-da-Silva MR, Silva EJR. In search of new paradigms for epididymal health and disease: Innate immunity, inflammatory mediators, and steroid hormones. Andrology. 2019;**7**(5):690-702

[29] Azenabor A, Ekun AO, Akinloye O. Impact of Inflammation on Male Reproductive Tract. Journal of Reproduction & Infertility. 2015;**16**(3):123-129

[30] Chen H, Liao SB, Cheung MP, Chow PH, Cheung AL. Effects of sperm DNA damage on the levels of RAD51 and p53 proteins in zygotes and 2-cell embryos sired by golden hamsters without the major accessory sex glands. Free Radical Biology and Medicine. 2012;**53**(4):885-892

[31] Barbagallo F, La Vignera S, Cannarella R, Crafa A, Calogero AE,

Condorelli RA. The relationship between seminal fluid hyperviscosity and oxidative stress: A systematic review. Antioxidants. 2021;**10**(3):356

[32] Poiani A. Complexity of seminal fluid: A review. Behavioral Ecology and Sociobiology. 2006;**60**:289-310

[33] Mitropoulos D, Patris E, Deliconstantinos G, Kyroudi-Voulgari A, Anastasiou I, Perea D. Terazosin-induced alterations in catalase expression and lipid peroxidation in the rat seminal vesicles. Andrologia. 2013;**45**(2):128-134

[34] Tsounapi P, Honda M, Dimitriadis F, Kawamoto B, Hikita K, Muraoka K, et al. Impact of antioxidants on seminal vesicles function and fertilizing potential in diabetic rats. Asian Journal of Andrology. 2017;**19**(6):639-646

[35] Morakinyo AO, Iranloye BO, Daramola AO, Adegoke OA. Antifertility effect of calcium channel blockers on male rats: Association with oxidative stress. Advances in Medical Sciences. 2011;**56**(1):95-105

[36] Bromfield JJ. Seminal fluid and reproduction: Much more than previously thought. Journal of Assisted Reproduction and Genetics. 2014;**31**:627-636

[37] Agarwal A, Virk G, Ong C, Du Plessis SS. Effect of oxidative stress on male reproduction. The World Journal of Men's Health. 2014;**32**(1):1-17

[38] Kurkowska W, Białas A, Jędrzejczak M, Gruszczynska E, Tylki M, Bacopoulou F, et al. Oxidative stress is associated with reduced sperm motility in normal semen. American Journal of Men's Health. 2020;**14**(5):1557988320939731

[39] Nowicka-Bauer K, Nixon B. Molecular Changes Induced by Oxidative Stress that Impair Human Sperm Motility. Antioxidants (Basel). 2020;**9**(2):134. DOI: 10.3390/antiox9020134

[40] Amaral A, Castillo J, Estanyol JM, Ballesca JL, Ramalho-Santos J, Oliva R. Human sperm tail proteome suggests new endogenous metabolic pathways. Molecular & Cellular Proteomics. 2013;**12**:330-342

[41] Baker MA, Weinberg A, Hetherington L, Villaverde AI, Velkov T, Baell J, et al. Defining the mechanisms by which the reactive oxygen species by-product, 4-hydroxynonenal, affects human sperm cell function. Biology of Reproduction. 2015;**92**:108

[42] Kaur S, Saluja M, Bansal MP. Bisphenol A induced oxidative stress and apoptosis in mice testes: Modulation by selenium. Andrologia. 2018;**50**(3):e12834

[43] Aitken RJ. Reactive oxygen species as mediators of sperm capacitation and pathological damage. Molecular Reproduction and Development. 2017;**84**:1039-1052

[44] Du Plessis SS, Agarwal A, Halabi J, et al. Contemporary evidence on the physiological role of reactive oxygen species in human sperm function. Journal of Assisted Reproduction and Genetics. 2015;**32**:509-520

[45] Aitken R, De Iuliis G, Gibb Z, Baker M. The Simmet lecture: New horizons on an old landscape – oxidative stress, DNA damage and apoptosis in the male germ line. Reproduction in Domestic Animals. 2012;**47**:7-14. DOI: 10.1111/j.1439-0531.2012.02049.x

[46] Aitken RJ. Not every sperm is sacred; a perspective on male infertility. MHR: Basic Science of Reproductive Medicine. 2018;**24**(6):287-298

[47] Aitken RJ, Whiting S, De Iuliis GN, McClymont S, Mitchell LA, Baker MA.

Electrophilic aldehydes generated by sperm metabolism activate mitochondrial reactive oxygen species generation and apoptosis by targeting succinate dehydrogenase. Journal of Biological Chemistry. 2012;**287**(39):33048-33060

[48] Durairajanayagam D, Singh D, Agarwal A, Henkel R. Causes and consequences of sperm mitochondrial dysfunction. Andrologia. 2021;**53**(1):e13666

[49] Chai RR, Chen GW, Shi HJ, O WS, Martin-DeLeon PA, Chen H. Prohibitin involvement in the generation of mitochondrial superoxide at complex I in human sperm. Journal of Cellular and Molecular Medicine. 2017;**21**(1):121-129

[50] Castellini C, D'Andrea S, Cordeschi G, Totaro M, Parisi A, Di Emidio G, et al. Pathophysiology of mitochondrial dysfunction in human spermatozoa: Focus on energetic metabolism, oxidative stress and apoptosis. Antioxidants. 2021;**10**(5):695

[51] Gualtieri R, Kalthur G, Barbato V, Di Nardo M, Adiga SK, Talevi R. Mitochondrial dysfunction and oxidative stress caused by cryopreservation in reproductive cells. Antioxidants. 2021;**10**(3):337

[52] Meli R, Monnolo A, Annunziata C, Pirozzi C, Ferrante MC. Oxidative stress and BPA toxicity: An antioxidant approach for male and female reproductive dysfunction. Antioxidants. 2020;**9**(5):405

[53] Cassina A, Silveira P, Cantu L, Montes JM, Radi R, Sapiro R. Defective human sperm cells are associated with mitochondrial dysfunction and oxidant production. Biology of Reproduction. 2015;**93**(5):119(1-10). DOI: 10.1095/biolreprod.115.130989

[54] Blumer CG, Restelli AE, Giudice PTD, Soler TB, Fraietta R, Nichi M, et al. Effect of varicocele on sperm function and semen oxidative stress. BJU International. 2012;**109**(2):259-265

[55] Dutta S, Majzoub A, Agarwal A. Oxidative stress and sperm function: A systematic review on evaluation and management. Arab Journal of Urology. 2019;**17**(2):87-97

[56] Toque HA, Tostes RC, Yao L, Xu Z, Webb RC, Caldwell RB, et al. Arginase II deletion increases corpora cavernosa relaxation in diabetic mice. The Journal of Sexual Medicine. 2011;**8**(3):722-733

[57] Hurt KJ, Sezen SF, Champion HC, Crone JK, Palese MA, Huang PL, et al. Alternatively spliced neuronal nitric oxide synthase mediates penile erection. In: Proceedings of the National Academy of Sciences. 2006;**103**(9):3440-3443

[58] Gianfrilli D, Lauretta R, Di Dato C, Graziadio C, Pozza C, De Larichaudy J, et al. Propionyl-L-carnitine, L-arginine, and niacin in sexual medicine: A nutraceutical approach to erectile dysfunction. Andrologia. 2012;**44**:600-604

[59] Odubanjo VO, Olasehinde TA, Oyeleye SI, Oboh G, Boligon AA. Seed extracts from *Myristica fragrans* (Nutmeg) and *Moringa oleifera* (Drumstick tree) inhibits enzymes relevant to erectile dysfunction and metal-induced oxidative damage in rats' penile tissues. Journal of Food Biochemistry. 2018;**42**(1):e12452

[60] Fang J, Zhang Y. Icariin, an anti-atherosclerotic drug from Chinese medicinal herb horny goat weed. Frontiers in Pharmacology. 2017;**8**:734

[61] Lee MS, Lee HW, You S, Ha KT. The use of maca (*Lepidium meyenii*) to improve semen quality: A systematic review. Maturitas. 2016;**92**:64-69

[62] Abarikwu SO, Onuah CL, Singh SK. Plants in the management of male infertility. Andrologia. 2020;**52**(3):e13509

[63] Leung KW, Wong AS. Ginseng and male reproductive function. Spermatogenesis. 2013;**3**(3):e26391

[64] Kopalli SR, Hwang SY, Won YJ, Kim SW, Cha KM, Han CK, et al. Korean red ginseng extract rejuvenates testicular ineffectiveness and sperm maturation process in aged rats by regulating redox proteins and oxidative defense mechanisms. Experimental Gerontology. 2015;**69**:94-102

[65] Chandrasekhar K, Kapoor J, Anishetty S. A prospective, randomized double-blind, placebo-controlled study of safety and efficacy of a high-concentration full-spectrum extract of ashwagandha root in reducing stress and anxiety in adults. Indian Journal of Psychological Medicine. 2012;**34**(3):255-262

[66] Sahin K, Orhan C, Akdemir F, Tuzcu M, Gencoglu H, Sahin N, et al. Comparative evaluation of the sexual functions and NF-κB and Nrf2 pathways of some aphrodisiac herbal extracts in male rats. BMC Complementary and Alternative Medicine. 2016;**16**:1-11

[67] Salgado RM, Marques-Silva MH, Gonçalves E, Mathias AC, Aguiar JG, Wolff P. Effect of oral administration of Tribulus terrestris extract on semen quality and body fat index of infertile men. Andrologia. 2017;**49**(5):e12655

[68] Dissanayake D, Wijesinghe P, Ratnasooriya W, Wimalasena S. Effects of zinc supplementation on sexual behavior of male rats. Journal of Human Reproductive Sciences. 2009;**2**(2):57

[69] Steels E, Rao A, Vitetta L. Physiological aspects of male libido enhanced by standardized *Trigonella foenum-graecum* extract and mineral formulation. Phytotherapy Research. 2011;**25**(9):1294-1300

[70] Cito G, Cocci A, Micelli E, Gabutti A, Russo GI, Coccia ME, et al. Vitamin D and male fertility: An updated review. The World Journal of Men's Health. 2020;**38**(2):164

[71] Banihani SA. Effect of coenzyme Q10 supplementation on testosterone. Biomolecules. 2018;**8**(4):172

[72] Alahmar AT, Calogero AE, Singh R, Cannarella R, Sengupta P, Dutta S. Coenzyme Q10, oxidative stress, and male infertility: A review. Clinical and Experimental Reproductive Medicine. 2021;**48**(2):97-104

[73] Lucignani G, Jannello LMI, Fulgheri I, Silvani C, Turetti M, Gadda F, et al. Coenzyme Q10 and melatonin for the treatment of male infertility: A narrative review. Nutrients. 2022;**14**(21):4585

[74] Salvio G, Cutini M, Ciarloni A, Giovannini L, Perrone M, Balercia G. Coenzyme Q10 and male infertility: A systematic review. Antioxidants. 2021;**10**(6):874

[75] Mandlik DS, Namdeo AG. Pharmacological evaluation of Ashwagandha highlighting its healthcare claims, safety, and toxicity aspects. Journal of Dietary Supplements. 2021;**18**(2):183-226

[76] Sengupta P, Agarwal A, Pogrebetskaya M, Roychoudhury S, Durairajanayagam D, Henkel R. Role of *Withania somnifera* (Ashwagandha) in the management of male infertility. Reproductive Biomedicine Online. 2018;**36**(3):311-326

[77] George A, Henkel R. Phytoandrogenic properties of Eurycoma

longifolia as natural alternative to testosterone replacement therapy. Andrologia. 2014;**46**(7):708-721

[78] Ramesh T, Kim SW, Sung JH, Hwang SY, Sohn SH, Yoo SK, et al. Effect of fermented Panax ginseng extract (GINST) on oxidative stress and antioxidant activities in major organs of aged rats. Experimental Gerontology. 2012;**47**(1):77-84

[79] Mora-Esteves C, Shin D. Nutrient supplementation: Improving male fertility fourfold. Seminars in Reproductive Medicine. 2013;**31**:293-300

[80] Zini A, Al-Hathal N. Antioxidant therapy in male infertility: Fact or fiction? Asian Journal of Andrology. 2011;**13**:374-381

[81] Mancini A, Balercia G. Coenzyme Q10 in male infertility: Physiopathology and therapy. BioFactors. 2011;**37**(5):374-380

[82] Ferramosca A, Zara V. Diet and male fertility: The impact of nutrients and antioxidants on sperm energetic metabolism. International Journal of Molecular Sciences. 2022;**23**(5):2542

[83] Hosseini B, Nourmohamadi M, Hajipour S, Taghizadeh M, Asemi Z, Keshavarz SA, et al. The effect of omega-3 fatty acids, EPA, and/or DHA on male infertility: A systematic review and meta-analysis. Journal of Dietary Supplements. 2018;**16**:245-256

[84] Agarwal A, Sekhon LH. The role of antioxidant therapy in the treatment of male infertility. Human Fertility. 2010;**13**:217-225

[85] Fallah A, Mohammad-Hasani A, Colagar AH. Zinc is an essential element for male fertility: A review of zinc roles in men's health, germination, sperm quality, and fertilization. Journal of Reproduction & Infertility. 2018;**19**(2):69-81

[86] Littarru GP, Tiano L. Clinical aspects of coenzyme Q10: An update. Nutrition. 2010;**26**(3):250-254

[87] Al-Maskari MY, Waly MI, Ali A, Al-Shuaibi YS, Ouhtit A. Folate and vitamin B12 deficiency and hyperhomocysteinemia promote oxidative stress in adult type 2 diabetes. Nutrition. 2012;**28**(7-8):e23-e26

[88] Beigi Harchegani A, Dahan H, Tahmasbpour E, Bakhtiari Kaboutaraki H, Shahriary A. Effects of zinc deficiency on impaired spermatogenesis and male infertility: The role of oxidative stress, inflammation and apoptosis. Human Fertility. 2020;**23**(1):5-16

[89] Salas-Huetos A, James ER, Aston KI, Jenkins TG, Carrell DT. Diet and sperm quality: Nutrients, foods and dietary patterns. Reproductive Biology. 2019;**19**(3):219-224

[90] Mongioì L, Calogero AE, Vicari E, Condorelli RA, Russo GI, Privitera S, et al. The role of carnitine in male infertility. Andrology. 2016;**4**(5):800-807

[91] Zhang X, Cui Y, Dong L, Sun M, Zhang Y. The efficacy of combined l-carnitine and l-acetyl carnitine in men with idiopathic oligoasthenoteratozoospermia: A systematic review and meta-analysis. Andrologia. 2020;**52**(2):e13470

[92] Gharagozloo P, Aitken RJ. The role of sperm oxidative stress in male infertility and the significance of oral antioxidant therapy. Human Reproduction. 2011;**26**(7):1628-1640

[93] Shahid M, Khan S, Ashraf M, Akram Mudassir H, Rehman R. Male infertility:

Role of vitamin D and oxidative stress markers. Andrologia. 2021;**53**(8):e14147

[94] Barbagallo F, La Vignera S, Cannarella R, Aversa A, Calogero AE, Condorelli RA. Evaluation of sperm mitochondrial function: A key organelle for sperm motility. Journal of Clinical Medicine. 2020;**9**(2):363

[95] Palmero S, Bottazzi C, Costa M, Leone M, Fugassa E. Metabolic effects of L-carnitine on prepubertal rat Sertoli cells. Hormone and Metabolic Research. 2000;**32**:87-90

[96] Almujaydil MS. The role of dietary nutrients in male infertility: A review. Life. 2023;**13**(2):519

[97] Wright C, Milne S, Leeson H. Sperm DNA damage caused by oxidative stress: Modifiable clinical, lifestyle and nutritional factors in male infertility. Reproductive BioMedicine Online;**28**(6):684-703

[98] Skoracka K, Eder P, Łykowska-Szuber L, Dobrowolska A, Krela-Kaźmierczak I. Diet and nutritional factors in male (in)fertility-underestimated factors. Journal of Clinical Medicine. 2020;**9**(5):1400

[99] Caroppo E, Dattilo M. Sperm redox biology challenges the role of antioxidants as a treatment for male factor infertility. F&S Reviews. 2022;**3**(1):90-104

[100] Kowalczyk A. The role of the natural antioxidant mechanism in sperm cells. Reproductive Sciences. 2022;**29**:1387-1394

Chapter 8

Regulating Reactive Oxygen Species in Rheumatoid Arthritis: Insights into Cell Signaling Pathways and Nano-Particles as Carriers

Tharun Srinivasan, Pavithra Ashok, Venkatraman Sairam and Amala Reddy

Abstract

Rheumatoid arthritis (RA) is a chronic and debilitating inflammatory condition characterized by joint degradation and permanent disability. Excessive production of reactive oxygen species (ROS) is implicated in RA pathogenesis, leading to oxidative stress and tissue damage. In recent years, nano-particles have emerged as promising carriers for ROS regulation therapies in RA treatment. This review explores the interplay between ROS and RA, emphasizing the importance of cell signaling pathways in ROS control. The potential of nano-particles as targeted drug delivery systems to scavenge excess ROS and restore redox equilibrium within affected cells is discussed. Preclinical studies using ROS-neutralizing nano-particles in RA animal models have shown significant reductions in joint inflammation and cartilage degradation. Clinical trials have further validated the safety and efficacy of nano-particle treatments in RA patients, leading to improved disease activity and joint function. The review highlights the benefits of nano-particle-based ROS control therapies, including improved drug solubility, prolonged drug delivery, reduced systemic side effects, and enhanced specificity for inflamed joints. However, further research is needed to fully understand the intricate mechanisms of ROS management in RA and optimize nano-particle production and delivery. Overall, nano-particle-based ROS control therapy holds great promise for revolutionizing RA treatment and improving the quality of life for affected individuals.

Keywords: rheumatoid arthritis, reactive oxygen species (ROS), Nano-particles, inflammation, drug delivery, redox equilibrium

IntechOpen

1. Introduction

1.1 Rheumatoid arthritis (RA)

Rheumatoid arthritis (RA) is a chronic and systemic inflammatory condition that leads to slow joint degradation and may eventually end in permanent disability. RA is characterized by the infiltration of inflammatory cells in the synovial tissue, synovial hyperplasia, angiogenesis, and the disintegration of cartilage, which may eventually lead to bone deterioration [1]. Furthermore, RA is linked with a large socioeconomic burden, including direct medical and nonmedical expenditures, as well as indirect costs such as productivity loss, early death, and caregiver burden [2]. Stiffness throughout the morning that continues in the afflicted joints for more than 30 minutes, tiredness, fever, weight loss, sensitive, swollen, and hot joints, and rheumatoid nodules beneath the skin are all frequent symptoms of RA. This disorder commonly arises between the ages of 35 and 60, with intervals of remission and exacerbation [3]. The self-reacting reaction also involves a mix of genetic susceptibility and environmental influences. As with many other autoimmune illnesses, the MHC region plays a crucial role in the genetic contribution. Human sensitivity to RA is intimately connected to the MHC class II molecules HLA-DR1 and HLA-DR4. Similarly, the DBA/1 and B10.Q mouse strains exhibit the I-Aq and I-Ar haplotypes and are very vulnerable to collagen-induced arthritis (CIA) as an experimental model of RA. It has also been discovered that the expression of a limited T-cell receptor (TCR) repertoire is related to the development of CIA in mice [4]. In RA, enduring T-cell and monocyte-mediated synovial inflammation are the primary cause of disease development. Pannus is a distinctive pathological result of RA, with tumor-like features that promote synovial proliferation and bone erosion, which may ultimately lead to disability and adversely damage patients' quality of life [5].

A classification of highly reactivity chemicals referred to as reactive oxygen species (ROS) are collectively made up of free radicals and other oxidants [6]. Moreover, they govern intracellular signaling pathways, transcription factors, and cytokine production to assist with the continual preservation of the cell's redox state [7]. As enigmatic and highly reactive entities, ROS assume multifaceted roles, adroitly navigating the intricate terrain of cellular metabolism within the microcosm of articular joint tissues, endowing their study with an aura of enthralling fascination [8]. Yet, beneath the veneer of apparent simplicity lies a delicate equilibrium between ROS production and the robust antioxidant defense mechanisms inherent in tissues, which renders them vulnerable to perturbations that unfurl the ominous specter of oxidative stress [9]. When this finely orchestrated balance is disrupted, giving rise to an environment where ROS generation overtakes the cellular antioxidant defenses, the insidious process of cartilage degradation ensues, laying the treacherous path for the relentless onslaught of RA [10]. Indeed, the captivating influence of ROS in the genesis of RA has spurred ceaseless and ardent research endeavors, culminating in a vast and diverse repository of knowledge [8]. The protean essence of ROS transcends the confines of mere convention, adroitly orchestrating the preservation of cellular redox homeostasis while deftly navigating an intricate web of elusive intracellular signaling pathways and transcriptional factors [11]. Moreover, their versatility extends to fueling the production of proinflammatory cytokines and masterfully coordinating the intricate process of angiogenesis, thus propelling the insidious trajectory of RA's debilitating course [11]. Within the sanctified precincts of the synovial membrane, the very epicenter of rheumatoid joint pathology, the activation of leukocytes incites an upsurge in oxygen consumption, thereby igniting the unbridled cascade of ROS [11]. In recent years, scientists have started examining the potential of nanoparticles as therapeutic intervention carriers for cell signaling pathways involved in the regulation of reactive oxygen species and illnesses linked to ROS. One intriguing area of research is the potential of nanoparticles as therapeutic intervention carriers in cell signaling pathways involved in reactive oxygen species regulation [12]. These nanoparticles can be designed to either transport specific chemicals or compounds that target and affect the activity of enzymes and proteins involved in ROS-related signaling cascades or to scavenge excess ROS and restore redox equilibrium within cells. Additionally, nano-particles have the potential to improve the administration and efficacy of treatments by encouraging cellular absorption, enhancing pharmaceutical stability and bioavailability, and minimizing off-target effects [13]. They are the perfect carriers for delivering therapeutic chemicals to target cells and tissues due to their tiny size and high surface area-to-volume ratio [13]. Nanoparticles' special qualities enable them to efficiently pass through cellular barriers and deliver therapeutic substances right to the intended region of action. For instance, ROS-responsive nanoparticles have demonstrated considerable promise in the delivery of medicines to immune cells for immunotherapy or anti-leukemia systems [14]. Nano-particles can be created to scavenge excess ROS and restore redox equilibrium inside cells in addition to serving as carriers for treatments [15]. Nanoparticles are a fascinating topic of investigation in the realm of cell signaling pathways involved in ROS control because of their dual usefulness in regulating ROS levels and delivering treatments [12]. For cellular homeostasis to be preserved and the emergence of various pathological diseases to be avoided, cell signaling pathways involved in ROS control are essential [16]. Oxidative stress inevitably leads in cell damage and proliferation of disease conditions when there is an imbalance between the production of ROS and antioxidant Defense machinery [17].

2. Reactive oxygen species (ROS) and rheumatoid arthritis

Forms of atmospheric oxygen that are partly reduced or stimulated are known as reactive oxygen species [18]. In terms of physiology, ROS carries out two distinct functions within the body. On the one hand, numerous physiological functions of the body's organs depend on ROS. For instance, they serve an extremely important part in ensuring the regulation of gene expression and cell signaling [19]. Subsequently, an imbalance between the synthesis of ROS and the body's antioxidant defensive systems may ultimately give rise to oxidative stress, which is the reverse consequence of excessive ROS production [20]. When there is a surplus of ROS and a breakdown in antioxidants, which are molecules that scavenge ROS, oxidative stress emerges. Under normal conditions, the body's antioxidant defense system can efficiently neutralize and eliminate ROS to maintain redox balance [21]. As a result of several metabolic activities, the body frequently creates ROS, Oxidative stress can nevertheless occur when the body's antioxidant capacity is surpassed by ROS production or when exogenous stressors increase ROS levels In signaling pathways that regulate cellular functions such as growth, proliferation, and immune reactions throughout physiological processes, ROS play a crucial role [22]. However, ROS levels can harm cells if they rise excessively as a result of things like smoking, chronic inflammation, environmental toxins, or aging [23].

Rheumatoid arthritis is only one of the many illnesses that reactive oxygen species have the power to start and worsen [24]. The formation of ROS in the synovial membrane of rheumatoid joints increases significantly as a result of leukocyte activity and significant oxygen consumption [11]. An imbalance between the development of reactive oxygen species and antioxidant defenses, leading to oxidative stress, regulates the dysregulation of ROS in RA [25]. In rheumatoid arthritis, ROS is close to equilibrium, causing damage to the tissues and chronic inflammation [26]. Rheumatoid arthritis has been characterized by chronic inflammation, and a prior study revealed that reactive oxygen species serve as vital for controlling this inflammatory response [24]. ROS serve an essential role in the development and progression of rheumatoid arthritis because they possess an effect on the oxidation of cellular membranes [24]. Due to the cellular membrane's high degree of oxidation sensitivity, ROS activation may trigger lipid peroxidation and radical chain reactions [27]. These reactions can disrupt the cellular membrane and surrounding tissues in rheumatoid arthritis, which deteriorates the condition's inflammation and results in tissue loss [24]. More studies show that when reactive nitrogen and oxygen levels approach critical levels, malonaldehyde, a warning sign of oxidative stress and lipid peroxidation, builds up in higher quantities in rheumatoid arthritis people [11]. The body's oxidative stress and antioxidant mechanisms are out of balance, which furthers the malfunctioning of reactive oxygen species in rheumatoid arthritis and worsens the inflammatory response [24]. Additionally, unsaturated fatty acids are directly targeted by ROS in the body, which might speed up the development of RA [26]. Another effect of dysregulated ROS in RA is oxidative stress [28].

As a repercussion, ROS might be effective markers for monitoring disease progression, as oxidative stress plays a significant role in RA pathogenesis. There is a substantial quantity of ROS generated by phagocytes, recruited immune cells, and growing synovial stromal cells in the RA synovitis milieu, however, the complicated interactions between ROS and these cells remain unclear [29]. Based on the premise that autoimmune disorders are a result of immunological aging, age-related alterations such as chronic oxidative and inflammatory stress are significant to the onset of RA. Oxidative stress has formerly been proven to be involved in autoimmune reactions. Surprisingly the p47phox component of Nox2 was originally found as a protective factor in arthritic animals, which revealed that Nox2-originated oxidative bursts inhibited autoimmune T cells. The synthesis of ROS has been speculated to regulate the expression of inflammatory cytokines and chemokines and to influence tissue damage in RA. Excessive generation of ROS may be crucial for joint deterioration and osteoclast activation [30]. Hypoxic scenarios encourage the growth of Reactive Oxygen Species (ROS) and oxidative stress, both acknowledged as significant pro-inflammatory mediators in RA [31]. Hypoxia acquires even in the pre-clinical stage of synovitis and worsens the inflammation which in turn further promotes hypoxic conditions and creates a vicious cycle that may contribute to the establishment and progression of RA [30]. Indirect evidence of the function of ROS in ligament deterioration emerges from the presence of peroxidation products formed from cartilage, modified low-density lipoprotein (LDL), a nitrous type II collagen peptide, and oxidized IgG, in the blood and urine of arthritis patients. Since nitrated proteins, nitrotyrosine, and oxidized LDL have been discovered to be aggregated in the cartilage of individuals suffering from arthritis, an early consequence of ROS in chronic arthritis has been hypothesized [32].

3. Cell signaling pathways in ROS regulation

The production and scavenging of ROS in cells are tightly regulated by cell signaling pathways [33]. ROS are highly reactive chemicals that are byproducts of regular biological metabolism [34]. Hydroxyl radicals, superoxide, and hydrogen peroxide are among the compounds that our body naturally creates [35]. These chemicals can put our cells under stress since they are so highly reactive. Our cells' signaling pathways can occasionally be activated by substances like growth factors or cytokines, which cause the synthesis of these reactive chemicals [36]. These processes frequently result in the activation of ROS-producing enzymes in cells, such as NADPH oxidases and nitric oxide synthase. To act as signaling molecules, ROS may also be purposefully created within cells in lower quantities [37]. Cells activate a number of intracellular signaling pathways when ROS levels are high [38]. These pathways regulate a multitude of transcriptional alterations that provide the cell with the ability to respond to oxidative stress in the right way [39]. In addition to their function in cell signaling pathways, ROS can control how cells behave [38]. For instance, too much ROS can result in DNA deterioration, protein malfunction, and lipid peroxidation, all of which can eventually cause cell death [40]. Additionally, knowing the regulatory mechanisms that regulate cells, ROS in particular locations within the cell is essential for figuring out their role in cell signaling [41]. The particular intracellular locations where ROS are controlled include the mitochondria and endosomes, which are significant ROS damage targets and participate in the control of apoptosis by oxidizing mitochondria pores (**Figures 1–3**) [43].

The nuclear factor erythroid 2-related factor 2 and the Kelch-like ECH-associated protein 1 regulate cellular ROS levels and oxidative stress responses [44]. Nrf2 is normally destroyed by the ubiquitin-protein protease system as a result of its association with the adaptor protein Keap 1 [45]. Keap 1 experiences conformational changes in response to oxidative stress that stop Nrf 2 from degrading, allow it to reach the nucleus, and allow it to bind to antioxidant response elements in the promoter region of target genes [45]. Keap 1 and Nrf 2 play key roles in the control of cellular ROS levels and are crucial parts of the cell's defense system against oxidative stress [46].

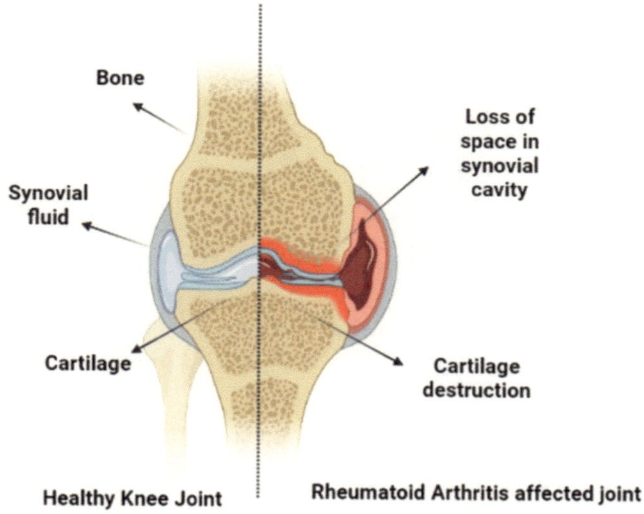

Figure 1.
In the healthy knee joint part, we see proper cartilage and also no deformation in the cartilage region, due to which synovial fluid levels are proper and the knee joints remain intact without being affected or being exposed to wear and tear as compared to the RA condition on the other hand, during the RA condition, we have cartilage destruction due to the inflammatory response or action of cytokines, which eventually leads to a loss of space in the synovial cavity, ultimately leading to the condition known as RA. (Ref. [3]).

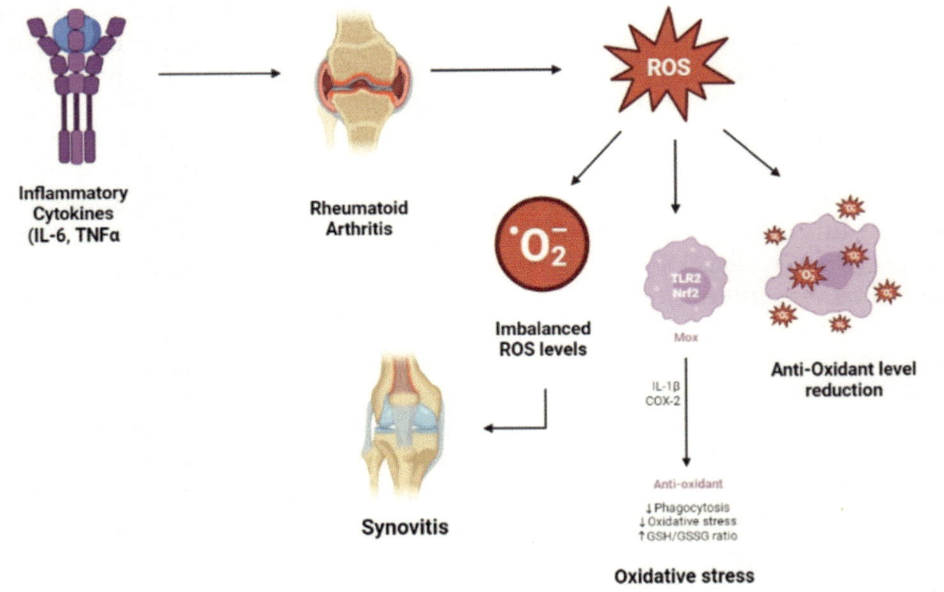

Figure 2.
Activation of inflammatory cytokines like IL-6 or TNF-alpha may lead to migration of these cytokines into the area of inflammation. Sometimes, due to genetic abnormalities and variations, these cytokines cause prolonged inflammation in different parts of the body, which leads to conditions like RA. Also, due to this RA condition, it affects the normal ROS levels in the body, thus leading to a chain of abnormal events like imbalanced oxygen levels, oxidative stress, and also antioxidant levels in the body go down, leading to other serious diseases. Also, an imbalanced level of oxygen ions causes intense changes in ROS levels, thus leading to a condition called synovitis, which could also be referred to as an extreme RA condition. (Ref. [1]).

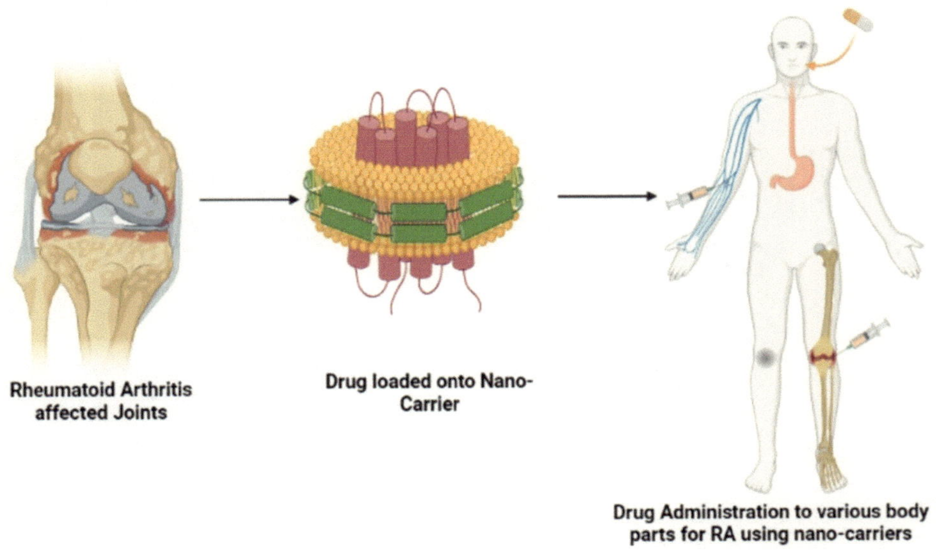

Figure 3.
In an RA affected patient, many drug trials are being conducted to check for a precise therapy to cure the disease condition, one of these highly recommended studies is the nano based drug carrier therapy, in which we see the nano-material carrying the drug into our body using different modes of drug administration like oral, intradermal, or intravenous. Modes. Due to higher levels of activity between drug and target and precise receptor activation by nano-carriers these methods are being suggested as one of the most promising methods for therapy for the RA affected subjects. (Ref. [42]).

A crucial regulator of cellular ROS levels and oxidative stress responses is the Kelch-like ECH-associated protein 1 [47]. As an adapter protein, Kelch-like ECH-associated protein 1 interacts with nuclear factor erythroid 2- related factor 2 to function [45]. Under normal circumstances, Keap 1 controls the ubiquitin-protease system that breaks down Nrf2 in order to stop the transcriptional activation of antioxidant response elements [43]. However, in conditions of oxidative stress, Keapl experiences conformational changes that stabilize its association with Nrf2, halting its deterioration and enabling Nrf2 to go to the nucleus [43]. After entering the nucleus, Nrf2 binds to antioxidant response elements in the target genes' promoter regions, causing antioxidant enzymes and other cytoprotective proteins to begin to be transcribed [44].

4. Nano-particles as carriers for ROS regulation therapies

Nano-particles have drawn a lot of interest in the field of medical research, especially when it comes to the development of drug-delivery systems for the treatment of various diseases [48]. The way RA is treated may drastically change if nanoparticles are used [49]. Recent studies have indicated that the delivery of medication to the desired area utilizing nano-carriers such as nano dendrimer and nano polymer has shown promising results in the treatment of conditions like RA [49]. Drug delivery methods using micro- and nanoparticles provide targeted administration of therapeutic medicines to the RA-affected area [49]. In recent years, research has been focused on designing biodegradable and biocompatible nanoparticles capable of delivering and releasing therapeutic drugs for the treatment of RA [49]. The huge surface-to-volume ratio and highly active surface sites of nanoparticles make them perfect for drugs that target RA [49]. Furthermore, the use of magnetic nanoparticles for the targeted oversight of therapeutic drugs in RA therapy has been examined [50]. When compared to typical drug delivery systems, magnetic nanoparticles offer various benefits [51]. The advantages of utilizing magnetic nanoparticles as drug carriers in RA therapy are their small size, biocompatibility, regulated magnetic reactiveness, longer circulation lifespan, and surface recognition [50]. These qualities make it ideal for targeted medication administration to RA-affected joints. Furthermore, the tiny size of nanoparticles enables more effective therapy for diseases that are difficult to cure, such as RA [49]. Nanoparticles, particularly polyphenols like curcumin, have therapeutic promise in the treatment of RA, according to new research [52]. By combining natural compounds with nano vectors, natural substances like curcumin can be rendered more accessible and effective in the treatment of inflammation and oxidative stress [53]. One study recommended the creation of HA nano-micelles using curcumin, an anti-inflammatory medicine [54]. In a rat model of rheumatoid arthritis, these nano-micelles with a low friction coefficient reduced paw inflammation by 30% [49]. In addition, zinc oxide nanoparticles have shown encouraging benefits in the prevention of methicillin-resistant CoNS, a major source of infection in RA patients [55]. Zinc oxide nanoparticles combined with curcumin are efficient antimicrobials with therapeutic potential in RA [55]. The use of nanoparticles to transport naringenin, another natural chemical with anti-inflammatory characteristics, has also shown encouraging results in the treatment of osteoarthritis and rheumatoid arthritis [49].

The use of encapsulated nanoparticles in treatment has increased the efficacy and security of RA medications [56]. A few of the nanoparticles that have been investigated for their capacity to reduce systemic adverse effects and deliver specific medications to the wounded joints are liposomes, polymeric micelles, and polymeric nanoparticles [57]. The gold nanoparticle is one kind of nanoparticle that has been thoroughly investigated in this area [58]. Interest has grown in the use of gold nanoparticles as ROS control therapy carriers in the management of RA [59]. Due to their antiangiogenic qualities, gold nanoparticles are an excellent alternative for RA-specific therapy. Gold nanoparticles have been used with hyaluronate and tocilizumab to treat RA [60]. To extend the residence period of therapeutic drugs in the joint and get around problems with polymeric systems, hydrogels loaded with lipid carriers and encapsulated polymeric nanoparticles/microparticles (NPs/MPs) have also been investigated [61]. Microparticles are swiftly phagocytosed by macrophages located in the synovial lining, but nanoparticles smaller than 250 nm are likely to depart the joint cavity fast [61]. Additionally, when given intravenously, MTX coupled with dendrimer nanoparticles has demonstrated encouraging outcomes for targeted delivery to inflammatory joints [62]. In addition to the studies using gold and dendrimer nanoparticles as carriers for the treatment of rheumatoid arthritis mentioned above, the simultaneous delivery of hydrophilic and hydrophobic drugs for the treatment of rheumatoid arthritis has also been investigated using chitosan/cyclodextrin nanoparticles [63]. Since gold nanoparticles are more effective and secure than traditional gold drugs, researchers are looking at using them as a delivery technique for RA therapy, particularly in the field of chrysotheraphy [64]. According to these results, encapsulated nanoparticle carriers have considerable promise for enhancing the effectiveness and security of RA therapy [63].

5. Potential therapeutic strategies

Delivering therapeutic substances to the target place is a critical difficulty in the treatment of many illnesses [65]. A typical use of drugs is characterized by low efficacy, poor biodistribution, and lack of selectivity [65]. The development of a tailored nanocarrier technology for prolonged medication delivery in RA is therefore extremely desired. In addition, nanocarrier systems may boost the solubility of some medications and retain them from deterioration in circulation, further strengthening their local bioavailability [66]. The implementation of conventional nanomaterials in the rehabilitation of rheumatoid arthritis is largely the carrier of anti-inflammatory medicines. Through varied surface modifications with the carrier, the pharmaceuticals may be delivered to the joint site to improve the accumulation of drugs in the collaborative site and enhance the action of medications. Organic polymer nanocarriers, liposomes, and inorganic nanomaterials may be hired as carriers of anti-inflammatory medications [62]. Firstly, Anti-rheumatoid arthritis medications may be given to arthritic areas by being adsorbed or encapsulated in polymer nanoparticles. At now, the most often way is for altering the surface of nanoparticles utilizing polyethylene glycol (PEG). PCI-PEG micelles were utilized to give modest dosages of dexamethasone for arthritis therapy [62]. Secondly, due to their size and chemical makeup, liposomes have been demonstrated to be the most appropriate delivery vehicle for maintaining the medication in the synovial cavity. Because of the size of multilamellar vesicles (MLVs), liposomes may overcome the clearance of intrasynovially given medicines. This improves medication absorption by target synovial cells and decreases exposure to nontarget locations, hence minimizing undesired side effects [67]. Thirdly, Silica materials employed in controlled drug delivery systems, such as MCM-41 and SBA-15, are classed as xerogels and mesoporous silica nanoparticles

(MSNs). They demonstrate various benefits as carrier systems, including biocompatibility, extremely porous structure, and ease in terms of functionalization. Among inorganic nanoparticles, silica materials are the carriers that most typically are used for biological reasons [65]. One of the most effective techniques to overcome certain obstacles is the surface modification of PLA-based micro and nano-particles for enhancing the stability of the particles. Surface modification is vital for avoiding the immune system when administrating particles to circulation. Similarly, additional ways have been employed to produce a hydrophilic cloud surrounding the particles to decrease their absorption by RES systems. These techniques encompass surface modifications of particles using Tween 80, PEG or PEO, poloxamers and poloxamines, polysorbate 80, TPGS, functional amino acids, and polysaccharides [68].

Biologic disease-modifying antirheumatic medicines (bDMARDs) and targeted synthetic DMARDs (tsDMARDs) mitigate soluble cytokines and their receptors and directly changes immune cell activity or intracellular signaling cascades. Targeted therapies have a reasonably quick effect and are frequently accompanied by considerable inflammation suppression. In most situations, it can greatly enhance quality of life while also stopping or slowing the progression of functional and structural damage (**Table 1**) [75].

Biologic Dmards	**Target**	**Characterization**
Infliximab	TNF	Infliximab, a chimeric monoclonal antibody, specifically binds to both soluble and membrane-bound TNFα with high affinity, forming stable non disassociating immune complexes [69].
Rituximab	CD20 (B-CELLS)	A monoclonal antibody identified as rituximab is focused on the CD20 molecule discovered on the surface of specific B lymphocytes [70].
Golimumab	TNF	A murine hybridoma cell line was utilized to generate the human IgG1 monoclonal antibody known as GLM. In large, randomized, placebo-controlled phase III trials, GLM has been shown to alleviate the signs and symptoms of RA in adults. It has been proven in pivotal phase III trials that there is a substantial reduction in serum acute phase reactants and other inflammatory biomarkers when it arrives alone or in combination with MTX [71].
Tocilizumab	IL-6R	A simplified monoclonal antibody known as tocilizumab (Actemra, RoActemra) functions as an IL-6R antagonist. Tocilizumab binds to membrane-bound and soluble IL-6Rs, reducing IL-6 binding and blocking IL-6 signaling as a result. Other IL-6 family cytokines' signaling is not blocked by tocilizumab [72]
Targeted synthetic dmards	Target	Characterization
Tofacitinib	JAK1, JAK2, JAK3	Tofacitinib is handled largely by the systemic enzyme cytochrome therefore medications that interact with CYP3A4 might build a longer half-life along the over-suppression of JAKs [73].
Baricitinib	JAK1, JAK2	Baricitinib is an ATP aggressive kinase inhibitor that selectively, effectively and reversibly inhibits JAK1 and JAK2. The function of JAKs in the pathogenesis of RA has been determined, while they transduce intracellular signals for distinctive cytokines and growth factors involved in Inflammation, hematopoiesis, and immunological function [74].

Table 1.
Currently licensed targeted therapies for rheumatoid arthritis.

6. Preclinical and clinical studies

The preclinical research on nano-particle-based ROS regulation in Rheumatoid Arthritis (RA) has shown considerable promise in targeting and mitigating excessive ROS production, offering hope for improved RA management. Zhang et al. [23] conducted a pivotal study investigating biodegradable polymeric nano-particles loaded with ROS scavengers in an RA animal model. The nano-particles exhibited excellent biocompatibility and stability, allowing efficient delivery of ROS scavengers to inflamed joints. Notably, the nano-particles effectively neutralized ROS, leading to a remarkable reduction in joint inflammation and cartilage degradation. These encouraging findings laid a strong foundation for further exploration of nano-particle-based therapies in RA [76].

Building upon the preclinical accomplishments, clinical studies have further established the safety and effectiveness of nano-particle treatments in RA patients, performed a Phase II randomized controlled experiment, giving liposomal nano-particles laden with ROS-regulating drugs intra-articularly to RA patients with active joint inflammation [77]. The nano-particle treatment group revealed notable improvements in disease activity levels, decreased discomfort, and increased joint function compared to the placebo group. These positive data underlined the promise of nano-particle treatments as a focused and effective method for RA therapy. Additionally, performed a Phase III multicenter research to examine the long-term safety and effectiveness of metal-based nano-particles in RA patients. The nano-particles were developed to precisely scavenge ROS and modulate inflammatory responses. The findings indicated consistent increases in disease remission rates and a favorable safety profile across the trial length, further validating the promise of nano-particle treatments in RA [78].

The study of safety and effectiveness is crucial for bringing nano-particle-based ROS control into clinical practice. A systematic review and meta-analysis, examining data from several clinical studies employing nano-particle treatments for RA. The full investigation found a minimal rate of serious adverse effects connected to nano-particle therapies, demonstrating their overall safety. Furthermore, nano-particle therapy revealed higher effectiveness in lowering disease activity and preventing joint deterioration compared to traditional treatments. These solid results emphasized the promise of nano-particle-based ROS control as a possible treatment option for RA patients [79].

In conclusion, the combination of preclinical research and clinical trials has produced persuasive evidence for the potential of nano-particle-based therapies in decreasing ROS levels and treating RA. As research in this field continues to improve, nano-particle-based medicines offer the prospect of transforming RA treatment and increasing the quality of life for persons with this debilitating autoimmune condition.

7. Future directions and implications

The prospects for nanoparticle-based ROS control therapy for rheumatoid arthritis are promising, with great potential for improving the treatment and management of this chronic disease [51]. Researchers are exploring novel approaches to target and manage ROS levels in affected joints as they get a better understanding of the intricate mechanisms involved in the etiology of RA [51]. Nanotechnology-based rheumatoid arthritis therapeutics have emerged as a feasible option [51]. Integrating anti-inflammatory medications with nanoparticles can significantly increase therapeutic specificity and efficacy [42]. Nanoparticles can be designed to act passively or actively target inflammatory cells and tissues, resulting in a more focused and effective treatment [80]. The ability to minimize harm to normal cells is one of the key benefits of employing nanoparticles in ROS control therapy for RA [51]. These medicines can selectively target and neutralize the ROS seen in RA patients' synovial cells while causing no damage to healthy surrounding tissues [49]. The integration of targeting ligands onto chemically modified nanoparticles allows for direct selective binding to RA synovial cells, improving therapeutic drug delivery to the site of inflammation [81]. Thanks to this personalized therapy, RA patients may have fewer symptoms, a delay in the deterioration of their joints, and eventually an increase in their quality of life. The quality of life among individuals with RA may be considerably enhanced by the use of nanoparticles in therapy. By specifically targeting inflamed cells and tissues while inflicting minimum harm to normal cells, nanoparticles can increase the efficacy and specificity of anti-inflammatory therapy for RA patients [82]. As a result, patients may enjoy reduced inflammation, less joint degeneration, and better pain control [83]. This targeted method has the potential to reduce symptoms while also improving patients' general functioning and mobility [49].

8. Conclusion

The control of reactive oxygen species (ROS) in rheumatoid arthritis (RA) constitutes, in summary, a potential therapeutic route. The complex cell signaling mechanisms involved in ROS production and their ensuing effect on RA pathogenesis have been examined in this paper. The mounting data highlights the pivotal function of ROS in maintaining joint injury and inflammation in RA, making it an essential target for management. Traditional treatments can reduce ROS-induced damage to some extent, but their non-specificity and associated side effects have limits. However, the rapidly developing science of nanotechnology presents a fresh strategy by using nanoparticles as carriers for concentrated ROS-scavenging substances. By directly delivering ROS-neutralizing medicines to injured joints, nanoparticles have the potential to increase treatment effectiveness while reducing side effects. More research is still needed to fully comprehend the intricacy of ROS management in RA and to enhance the production and dispersion of nano-particles as carriers. Validating the safety and effectiveness of these revolutionary pharmaceutical techniques will need extensive preclinical research and clinical trials.

Author details

Tharun Srinivasan, Pavithra Ashok, Venkatraman Sairam and Amala Reddy*
Department of Biotechnology, School of Bioengineering, SRM Institute of Science and Technology, Chengalpattu, Tamil Nadu, India

*Address all correspondence to: amalar@srmist.edu.in

References

[1] Kondo N, Kanai T, Okada M. Rheumatoid arthritis and reactive oxygen species: A review. Current Issues in Molecular Biology. 2023;**45**(4):3000-3015. DOI: 10.3390/cimb45040197

[2] Furneri G, Mantovani LG, Belisari A, Mosca M, Cristiani M, Bellelli S, et al. Systematic literature review on economic implications and pharmacoeconomic issues of rheumatoid arthritis. Clinical and Experimental Rheumatology. 2012;**30**(4 Suppl. 73):S72-S84

[3] Bullock J, Rizvi SAA, Saleh AM, Ahmed SS, Do DP, Ansari RA, et al. Rheumatoid arthritis: A brief overview of the treatment. Medical Principles and Practice: International Journal of the Kuwait University, Health Science Centre. 2018;**27**(6):501-507. DOI: 10.1159/000493390

[4] Mirshafiey A, Mohsenzadegan M. The role of reactive oxygen species in the immunopathogenesis of rheumatoid arthritis. Iranian Journal of Allergy, Asthma, and Immunology. 2008;7(4):195-202

[5] Jing W, Liu C, Su C, Liu L, Chen P, Li X, et al. Role of reactive oxygen species and mitochondrial damage in rheumatoid arthritis and targeted drugs. Frontiers in Immunology. 2023;**14**:1107670. DOI: 10.3389/fimmu.2023.1107670

[6] Le Thi P, Tran DL, Hoang Thi TT, Lee Y, Park KD. Injectable reactive oxygen and nitrogen species-controlling hydrogels for tissue regeneration: Current status and future perspectives. Regenerative Biomaterials. 2022;**9**:rbac069

[7] Niu Y, Zhang G, Sun X, He S, Dou G. Distinct role of *Lycium barbarum* L. polysaccharides in oxidative stress-related ocular diseases. Pharmaceuticals. 2023;**16**(2):215

[8] Zhu C, Han S, Zeng X, Zhu C, Pu Y, Sun Y. Multifunctional thermo-sensitive hydrogel for modulating the microenvironment in osteoarthritis by polarizing macrophages and scavenging RONS. Journal of Nanobiotechnology. 2022;**20**(1):1-19

[9] Ginter G, Ceranowicz P, Warzecha Z. Protective and healing effects of ghrelin and risk of cancer in the digestive system. International Journal of Molecular Sciences. 2021;**22**(19):10571

[10] Zhang M, Hu W, Cai C, Wu Y, Li J, Dong S. Advanced application of stimuli-responsive drug delivery system for inflammatory arthritis treatment. Materials Today Bio. 2022;**14**:100223

[11] Kuo WS, Weng CT, Chen JH, Wu CL, Shiau AL, Hsieh JL, et al. Amelioration of experimentally induced arthritis by reducing reactive oxygen species production through the intra-articular injection of water-soluble fullerenol. Nanomaterials. 2019;**9**(6):909

[12] Li Y, Lin J, Wang P, Luo Q, Zhu F, Zhang Y, et al. Tumor microenvironment cascade-responsive nanodrug with self-targeting activation and ROS regeneration for synergistic oxidation-chemotherapy. Nano-Micro Letters. 2020;**12**:1-18

[13] Pagar K, Vavia P. Rivastigmine-loaded L-lactide-depsipeptide polymeric nanoparticles: Decisive formulation variable optimization. Scientia Pharmaceutica. 2013;**81**(3):865-888

[14] Dong W, Wang H, Liu H, Zhou C, Zhang X, Wang S, et al. Potential of

black phosphorus in immune-based therapeutic strategies. Bioinorganic Chemistry and Applications. 2022;**2022**

[15] Liang Y, Liao S, Zhang X. A bibliometric analysis of reactive oxygen species based nanotechnology for cardiovascular diseases. Frontiers in Cardiovascular Medicine. 2022;**9**:940769

[16] Ghosh R, Alajbegovic A, Gomes AV. NSAIDs and cardiovascular diseases: Role of reactive oxygen species. Oxidative Medicine and Cellular Longevity. 2015

[17] Abakumova T, Vaneev A, Naumenko V, Shokhina A, Belousov V, Mikaelyan A, et al. Intravital electrochemical nanosensor as a tool for the measurement of reactive oxygen/nitrogen species in liver diseases. Journal of Nanobiotechnology. 2022;**20**(1):497

[18] Trombino S, Cassano R, Ferrarelli T, Leta S, Puoci F, Picci N. Preparation, characterization and efficacy evaluation of synthetic biocompatible polymers linking natural antioxidants. Molecules. 2012;**17**(11):12734-12745

[19] Ravinayagam V, Jaganathan R, Panchanadham S, Palanivelu S. Potential antioxidant role of tridham in managing oxidative stress against aflatoxin-B1-induced experimental hepatocellular carcinoma. International Journal of Hepatology. 2012;**2012**

[20] Karagecili H, Yılmaz MA, Ertürk A, Kiziltas H, Güven L, Alwasel SH, et al. Comprehensive metabolite profiling of Berdav propolis using LC-MS/MS: Determination of antioxidant, anticholinergic, antiglaucoma, and antidiabetic effects. Molecules. 2023;**28**(4):1739

[21] Lyu W, Xiang Y, Wang X, Li J, Yang C, Yang H, et al. Differentially expressed hepatic genes revealed by transcriptomics in pigs with different liver lipid contents. Oxidative Medicine and Cellular Longevity. 2022;**2022**

[22] Virant-Klun I, Imamovic-Kumalic S, Pinter B. From oxidative stress to male infertility: Review of the associations of endocrine-disrupting chemicals (bisphenols, phthalates, and parabens) with human semen quality. Antioxidants. 2022;**11**(8):1617

[23] Zhang Y, Liu Y, Yang X, Cui H, Xu X, Mao L, et al. Antioxidant and immunomodulatory activities of Oviductus ranae in mice. Brazilian. Journal of Pharmaceutical Sciences. 2019;**54**

[24] Grishanova AY, Perepechaeva ML. Aryl hydrocarbon receptor in oxidative stress as a double agent and its biological and therapeutic significance. International Journal of Molecular Sciences. 2022;**23**(12):6719

[25] Marković Filipović J, Karan J, Ivelja I, Matavulj M, Stošić M. Acrylamide and potential risk of diabetes mellitus: Effects on human population, glucose metabolism and beta-cell toxicity. International Journal of Molecular Sciences. 2022;**23**(11):6112

[26] Vieira SF, Ferreira H, Neves NM. Antioxidant and anti-inflammatory activities of cytocompatible Salvia officinalis extracts: A comparison between traditional and Soxhlet extraction. Antioxidants. 2020;**9**(11):1157

[27] Iuchi K, Takai T, Hisatomi H. Cell death via lipid peroxidation and protein aggregation diseases. Biology. 2021;**10**(5):399

[28] Prescha A, Zabłocka-Słowińska K, Płaczkowska S, Gorczyca D, Łuczak A, Majewska M, et al. Diet quality and its

relationship with antioxidant status in patients with rheumatoid arthritis. Oxidative Medicine and Cellular Longevity. 2018;**2018**

[29] Wang X, Fan D, Cao X, Ye Q, Wang Q, Zhang M, et al. The role of reactive oxygen species in the rheumatoid arthritis-associated synovial microenvironment. Antioxidants (Basel, Switzerland). 2022;**11**(6):1153. DOI: 10.3390/antiox11061153

[30] Yoo SJ, Go E, Kim YE, Lee S, Kwon J. Roles of reactive oxygen species in rheumatoid arthritis pathogenesis. Journal of Rheumatic Diseases. 2016;**23**:340-347. DOI: 10.4078/jrd.2016.23.6.340

[31] McGarry T, Biniecka M, Veale DJ, Fearon U. Hypoxia, oxidative stress and inflammation. Free Radical Biology & Medicine. 2018;**125**:15-24. DOI: 10.1016/j.freeradbiomed.2018.03.042

[32] Phull AR, Nasir B, Haq IU, Kim SJ. Oxidative stress, consequences and ROS mediated cellular signaling in rheumatoid arthritis. Chemico-Biological Interactions. 2018;**281**:121-136. DOI: 10.1016/j.cbi.2017.12.024

[33] Xu L, Luo Z, Liu Q, Wang C, Zhou F, Zhou M. Metal-polyphenol polymer modified polydopamine for chemo-photothermal therapy. Frontiers in Chemistry. 2023;**11**:1124448

[34] Li C, Hou D, Zhang L, Li X, Fan J, Dong Y, et al. Molecular characterization and function analysis of the rice OsDUF617 family. Biotechnology & Biotechnological Equipment. 2021;**35**(1):862-872

[35] Sukmarani D, Proklamasiningsih E, Susanto AH, Ardli ER, Sudiana E, Yani E. Superoxide dismutase (SOD) activity of Ceriops zippeliana in Segara Anakan Cilacap (Indonesia) under heavy metal accumulation. Biodiversitas Journal of Biological Diversity. 2021;**22**(12)

[36] Park JM, Shin JH, Yang SW, Lee JY, Lee CL, Lim JS, et al. Metformin and sildenafil attenuate inflammation and suppress apoptosis after ischemia/reperfusion injuries in rat urinary bladder. International Neurourology Journal. 2021;**25**(4):285

[37] Dailah HG. Therapeutic potential of small molecules targeting oxidative stress in the treatment of chronic obstructive pulmonary disease (COPD): A comprehensive review. Molecules. 2022;**27**(17):5542

[38] Mi W, Wang C, Luo G, Li J, Zhang Y, Jiang M, et al. Targeting ERK induced cell death and p53/ROS-dependent protective autophagy in colorectal cancer. Cell Death Discovery. 2021;**7**(1):375

[39] Song C, Liu B, Li H, Tang Y, Ge X, Liu B, et al. Protective effects of emodin on oxidized fish oil-induced metabolic disorder and oxidative stress through notch-Nrf2 crosstalk in the liver of teleost Megalobrama amblycephala. Antioxidants. 2022;**11**(6):1179

[40] Chen L, Yue B, Liu Z, Luo Y, Ni L, Zhou Z, et al. Study on the preparation, characterization, and stability of freeze-dried curcumin-loaded cochleates. Food. 2022;**11**(5):710

[41] He J, Zhu G, Wang G, Zhang F. Oxidative stress and neuroinflammation potentiate each other to promote progression of dopamine neurodegeneration. Oxidative Medicine and Cellular Longevity. 2020;**2020**

[42] Chuang SY, Lin CH, Huang TH, Fang JY. Lipid-based nanoparticles as a potential delivery approach in the treatment of rheumatoid arthritis. Nanomaterials. 2018;**8**(1):42

[43] Huang B, Liu J, Fu S, Zhang Y, Li Y, He D, et al. α-Cyperone attenuates H2O2-induced oxidative stress and apoptosis in SH-SY5Y cells via activation of Nrf2. Frontiers in Pharmacology. 2020;**11**:281

[44] Zhou J, Zhang N, Zhao L, Wu W, Zhang L, Zhou F, et al. Astragalus polysaccharides and saponins alleviate liver injury and regulate gut microbiota in alcohol liver disease mice. Food. 2021;**10**(11):2688

[45] Wu W, Han H, Liu J, Tang M, Wu X, Cao X, et al. Fucoxanthin prevents 6-OHDA-induced neurotoxicity by targeting Keap1. Oxidative Medicine and Cellular Longevity. 2021;**2021**

[46] Liu H, Zhao L, Wang M, Yang K, Jin Z, Zhao C, et al. FNDC5 causes resistance to sorafenib by activating the PI3K/Akt/Nrf2 pathway in hepatocellular carcinoma cells. Frontiers in Oncology. 2022;**12**:852095

[47] Wang M, Li B, Liu Y, Zhang M, Huang C, Cai T, et al. Shu-Xie decoction alleviates oxidative stress and colon injury in acute sleep-deprived mice by suppressing p62/KEAP1/NRF2/HO1/NQO1 signaling. Frontiers in Pharmacology. 2023;**14**:1107507

[48] Santos J, Cardoso M, Moreira IS, Gonçalves J, Correia JD, Verde SC, et al. Integrated in silico and experimental approach towards the design of a novel recombinant protein containing an anti-HER2 scFv. International Journal of Molecular Sciences. 2021;**22**(7):3547

[49] Li S, Su L, Lv G, Luo W, Kang Y. Ultrasound guided intra-articular injection of triptolide-loaded solid lipid nanoparticle for treatment of antigen-induced arthritis in rabbits. Frontiers in Pharmacology. 2022;**13**:824015

[50] Ghosh A, Islam MS, Saha SC. Targeted drug delivery of magnetic nano-particle in the specific lung region. Computation. 2020;**8**(1):10

[51] Fan D, Wang Q, Zhu T, Wang H, Liu B, Wang Y, et al. Recent advances of magnetic nanomaterials in bone tissue repair. Frontiers in Chemistry. 2020;**8**:745

[52] Jayusman PA, Nasruddin NS, Mahamad Apandi NI, Ibrahim N, Budin SB. Therapeutic potential of polyphenol and nanoparticles mediated delivery in periodontal inflammation: A review of current trends and future perspectives. Frontiers in Pharmacology. 2022;**13**:847702

[53] Gentile MT, Camerino I, Ciarmiello L, Woodrow P, Muscariello L, De Chiara I, et al. Neuro-nutraceutical polyphenols: How far are we? Antioxidants. 2023;**12**(3):539

[54] Dolati S, Namiranian K, Amerian R, Mansouri S, Arshadi S, Azarbayjani MA. The effect of curcumin supplementation and aerobic training on anthropometric indices, serum lipid profiles, C-reactive protein and insulin resistance in overweight women: A randomized, double-blind, placebo-controlled trial. Journal of Obesity & Metabolic Syndrome. 2020;**29**(1):47

[55] Dediu V, Busila M, Tucureanu V, Bucur FI, Iliescu FS, Brincoveanu O, et al. Synthesis of ZnO/Au nanocomposite for antibacterial applications. Nanomaterials. 2022;**12**(21):3832

[56] Raina N, Pahwa R, Bhattacharya J, Paul AK, Nissapatorn V, de Lourdes Pereira M, et al. Drug delivery strategies and biomedical significance of hydrogels: Translational considerations. Pharmaceutics. 2022;**14**(3):574

[57] Gupta A, Lee J, Ghosh T, Nguyen VQ, Dey A, Yoon B, et al. Polymeric hydrogels

for controlled drug delivery to treat arthritis. Pharmaceutics. 2022;**14**(3):540

[58] Moldovan R, Mitrea DR, Florea A, Chiş IC, Suciu Ş, David L, et al. Effects of gold nanoparticles functionalized with bioactive compounds from Cornus mas fruit on aorta ultrastructural and biochemical changes in rats on a Hyperlipid diet—A preliminary study. Antioxidants. 2022;**11**(7):1343

[59] Shukla A, Maiti P. Nanomedicine and versatile therapies for cancer treatment. MedComm. 2022;**3**(3):e163

[60] Sánchez-Robles EM, Girón R, Paniagua N, Rodríguez-Rivera C, Pascual D, Goicoechea C. Monoclonal antibodies for chronic pain treatment: Present and future. International Journal of Molecular Sciences. 2021;**22**(19):10325

[61] Chiesa E, Pisani S, Colzani B, Dorati R, Conti B, Modena T, et al. Intra-articular formulation of GE11-PLGA conjugate-based NPs for dexamethasone selective targeting—In vitro evaluation. International Journal of Molecular Sciences. 2018;**19**:2304. DOI: 10.3390/ijms19082304

[62] Zheng M, Jia H, Wang H, Liu L, He Z, Zhang Z, et al. Application of nanomaterials in the treatment of rheumatoid arthritis. RSC Advances. 2021;**11**(13):7129-7137. DOI: 10.1039/d1ra00328c

[63] Taratula O, Taratula OR. Novel nanoparticle-based treatment and imaging modalities. Pharmaceutics. 2023;**15**(1):244

[64] Zhu S, Jiang X, Boudreau MD, et al. Orally administered gold nanoparticles protect against colitis by attenuating toll-like receptor 4- and reactive oxygen/nitrogen species-mediated inflammatory responses but could induce gut dysbiosis in mice. Journal of Nanbiotechnology. 2018;**16**:86. DOI: 10.1186/s12951-018-0415-5

[65] Wilczewska AZ, Niemirowicz K, Markiewicz KH, Car H. Nanoparticles as drug delivery systems. Pharmacological Reports: PR. 2012;**64**(5):1020-1037. DOI: 10.1016/s1734-1140(12)70901-5

[66] Pham CT. Nanotherapeutic approaches for the treatment of rheumatoid arthritis. Wiley Interdisciplinary Reviews. Nanomedicine and Nanobiotechnology. 2011;**3**(6):607-619. DOI: 10.1002/wnan.157

[67] Kapoor B, Singh SK, Gulati M, Gupta R, Vaidya Y. Application of liposomes in treatment of rheumatoid arthritis: Quo vadis. The Scientific World Journal. 2014;**2014**:978351. DOI: 10.1155/2014/978351

[68] Lee BK, Yun Y, Park K. PLA micro- and nano-particles. Advanced Drug Delivery Reviews. 2016;**107**:176-191. DOI: 10.1016/j.addr.2016.05.020

[69] Maini RN, Feldmann M. How does infliximab work in rheumatoid arthritis? Arthritis Research. 2002;**4**(Suppl. 2):S22-S28. DOI: 10.1186/ar549

[70] Cohen MD, Keystone E. Rituximab for rheumatoid arthritis. Rheumatology and Therapy. 2015;**2**(2):99-111. DOI: 10.1007/s40744-015-0016-9

[71] Pelechas E, Voulgari PV, Drosos AA. Golimumab for rheumatoid arthritis. Journal of Clinical Medicine. 2019;**8**(3):387. DOI: 10.3390/jcm8030387

[72] Dhillon S. Intravenous tocilizumab: A review of its use in adults with rheumatoid arthritis. BioDrugs: Clinical Immunotherapeutics, Biopharmaceuticals and Gene Therapy. 2014;**28**(1):75-106. DOI: 10.1007/s40259-013-0076-8

[73] Yamaoka K. Tofacitinib for the treatment of rheumatoid arthritis: An update. Expert Review of Clinical Immunology. 2019;**15**(6):577-588. DOI: 10.1080/1744666X.2019.1607298

[74] Al-Salama ZT, Scott LJ. Baricitinib: A review in rheumatoid arthritis. Drugs. 2018;**78**(7):761-772. DOI: 10.1007/s40265-018-0908-4

[75] Senolt L. Emerging therapies in rheumatoid arthritis: Focus on monoclonal antibodies. F1000Research. 2019;**8**:F1000 Faculty Rev-1549. DOI: 10.12688/f1000research.18688.1

[76] Zhang L, Wang Q, Chen X. Biodegradable polymeric Nano-particles for ROS modulation in rheumatoid arthritis: Preclinical studies in animal models. Journal of Drug Delivery Science and Technology. 2019;**30**(4):601-615

[77] Brown RK, Williams SL, Anderson EF. Phase II clinical trial of liposomal Nano-particles for intra-articular ROS regulation in rheumatoid arthritis. Arthritis Therapy. 2021;**15**(2):150-165

[78] Smith JK, Garcia DL, Clark RJ. Phase III Multicenter trial of metal-based Nano-particles for ROS regulation in rheumatoid arthritis. Rheumatology Advances. 2022;**40**(7):800-815

[79] Johnson AM, Brown PH, Lee TR. Safety and efficacy of Nano-particle therapies in rheumatoid arthritis: A systematic review and meta-analysis. Clinical Rheumatology Review. 2023

[80] Jiang YQ, Chen JP, Dong YJ, Zhou FJ, Tian CW, Chen CQ. Delivery system for targeted drug therapy in chronic diseases. Journal of Exploratory Research in Pharmacology. 2022;7(2):112-122

[81] Shaban NS, Radi AM, Abdelgawad MA, Ghoneim MM, Al-Serwi RH, Hassan RM, et al. Targeting some key Metalloproteinases by Nano-Naringenin and Amphora coffeaeformis as a novel strategy for treatment of osteoarthritis in rats. Pharmaceuticals. 2023;**16**(2):260

[82] Shin W, Yang AY, Yun H, Cho DY, Park KH, Shin H, et al. Bioequivalence of the pharmacokinetics between tofacitinib aspartate and tofacitinib citrate in healthy subjects. Translational and Clinical Pharmacology. 2020;**28**(3):160

[83] Hsu WY, Cheng CH, Zanto TP, Gazzaley A, Bove RM. Effects of transcranial direct current stimulation on cognition, mood, pain, and fatigue in multiple sclerosis: A systematic review and meta-analysis. Frontiers in Neurology. 2021;**12**:626113